SPIDERS IN RICE
A Research Review

THE AUTHORS

S. Amandeep Singh is presently working as Assistant Professor at Khalsa College, Gardhiwala, Hoshiarpur, India. He did his B.Sc. (Agriculture) from Khalsa College Amritsar and went for M.Sc. (Entomology) at Punjab Agriculture University, Ludhiana. His field of research interest is apiculture such as bee behavior, bee pathology, bee toxicology, bee parasitology, bee ecology etc. He has a good teaching experience and guided five M.Sc. entomology students as an advisor for their research work. Moreover, he has a number of publications like research papers, book chapter and abstracts in national and international journals.

S. Randeep Singh is currently working as Assistant Professor and In-Charge, Crop Protection Division, PG Department of Agriculture of Khalsa College Amritsar. He did his M.Sc. (Zoology) and M.Phil (Zoology with specialization in entomology) from Department of Zoology, Guru Nanak Dev University, Amritsar in 2006 and 2009, respectively. His area of interest is Insect control, insect physiology and insect biochemistry. He has 8 years of teaching and research experience. He has already guided four M.Sc. Agriculture (Entomology) students for their thesis work and currently guiding nine students. He has published many research articles and book chapters.

Dr Mandeep Kaur is presently working Assistant Professor at Kanya Maha Vidayala, Jalandhar. She did her Ph.D. in Zoology (with specialization in Entomology) in 2018 from Guru Nanak Dev University, Amritsar. She has 7 years experience in research and teaching. She has 6 publications in National and International reputed journals and books. She published 11 abstracts in national and international conferences. She got Best poster presentation award at 4th International Science Congress.

ISBN: 9789390371990 (Int. Edition)

Publisher's Note:

Every possible effort has been made to ensure that the information contained in this book is accurate at the time of going to press, and the publisher and author cannot accept responsibility for any errors or omissions, however caused. No responsibility for loss or damage occasioned to any person acting, or refraining from action, as a result of the material in this publication can be accepted by the editor, the publisher or the author. The Publisher is not associated with any product or vendor mentioned in the book. The contents of this work are intended to further general scientific research, understanding and discussion only. Readers should consult with a specialist where appropriate.

Every effort has been made to trace the owners of copyright material used in this book, if any. The author and the publisher will be grateful for any omission brought to their notice for acknowledgement in the future editions of the book.

Published by : **Daya Publishing House®**
A Division of
Astral International Pvt. Ltd.
– ISO 9001:2015 Certified Company –
4736/23, Ansari Road, Darya Ganj
New Delhi-110 002
Ph. 011-43549197, 23278134
E-mail: info@astralint.com
Website: www.astralint.com

SPIDERS IN RICE
A Research Review

Editors
Amandeep Singh
Randeep Singh
Co-Editor
Mandeep Kaur

2020
Daya Publishing House®
A Division of
Astral International Pvt. Ltd.
New Delhi – 110 002

Preface

Anyone engaged in agriculture need to know about the role of an important biocontrol agent known as spiders that commonly affect pest populations in field crops. Pest control for the past several decades has relied heavily on the application of chemical insecticides. Valuable through these substances are, many problems have arisen due to their use, such as the appearance of resistant strains of pests and harmful effects on non-target organisms. Most entomologist has now agreed that priority must be given to the development of pest management systems which harmoniously integrate various control measures (including insecticides), if a pest control remain viable.

Spiders play an important role as a biological control agent in many cultivated plants. In fact, the latter usually provides the cornerstone of effective pest management systems. An important aspect therefore of entomology as an applied science is not only control of harmful insect species, but also the conservation and management of the environment.

Rice is the dominant staple food in the developing world and most essential energy source that provides food to about half of the world's population. The objective of this book is therefore to provide students, research scholars, practicing agriculturalists, agronomist and other interested persons with a basic introduction to various types of spiders in rice cropping systems. The treatment is introduced, not exhaustive, but should prove sufficient for understanding the basic role of spiders to manage the insect pest in rice cropping system. The foundation is also laid for further study in particular aspects for those that desire it. The present book is an outcome of the long need felt by the research scholars, scientists, students of Entomology and Zoology. A number of research manuscripts have been consulted in depth during the preparation of this manuscript that are listed in the references.

The book is divided into ten chapters containing seven tables and Six figures which make subjects more comprehensive and easily understandable. Chapter 1 is introductory and devoted to supplement the information of rice cropping systems and its new challenges. Chapter 2 deal with important insect pest in rice crop. Chapter 3 to 10 gives an overall idea about the spiders and in association with the rice cropping system and its effectiveness as an important biocontrol agent. Chapter 11 is aimed on describe the effect of pesticides on spiders.

We are thankful to almighty for this manuscript. We have received the help of different persons in various ways (encouragement, suggestions and blessings) who either have been our colleagues, learned teachers and research personals. We are also thankful to family members for their support and blessings

Contents

Contributors

Authors List

S. No.	Name	Address	Chapter wise Contribution
1.	S. Amandeep Singh	Assistant Professor, Department of Agriculture, Khalsa College, Garhdiwala Hoshiarpur	1-11
2.	S. Randeep Singh	Assistant Professor, Plant Protection Division, PG Department of Agriculture, Khalsa College Amritsar	1-4, 7, 9,11

Co-Authors List

S. No.	Name	Address	Chapter wise Contribution
1.	Dr. Mandeep Kaur	Assistant Professor , Department of Zoology, KMV College, Jalandhar	1, 4, 5, 6 and 8
2.	Ms. Kiranjot Kaur	Assistant Professor, Plant Protection Division, PG Department of Agriculture, Khalsa College Amritsar	4, 5 and 10
3.	Dr. Kapila Mahajan	Assistant Professor, Department of Zoology, DAV College, Jalandhar	2 and 8
4.	Dr. Gurbax Singh	Associate Professor , PG Department of Agriculture, Khalsa College Amritsar	5 and 6
5.	S. Amarinder Singh	Assistant Professor, PG Department of Agriculture, Khalsa College Amritsar	6
6.	S. Maninderjit Singh	Assistant Professor, PG Department of Agriculture, Khalsa College Amritsar	7
7.	Dr. Abhinay Thakur	Assistant Professor, Department of Zoology, DAV College, Jalandhar	8
8.	Dr. Ripudaman Parihar	Assistant Professor, Department of Zoology, DAV University, Jalandhar	9
10.	Dr. Jyoti Mahajan	Assistant Professor, Plant Protection Division, PG Department of Agriculture, Khalsa College Amritsar	11

Chapter 1

Rice Cropping System and its Challenges

Amandeep Singh, Randeep Singh and Mandeep Kaur

Rice is the dominant staple food in the developing world. This is the most essential energy source that provides food to about half of the world's population (IRRI, 1989). In many countries, rice production plays an important role in the economic development and any factor that decrease the production of this crop can adversely affect these countries. Almost in all rice producing countries of the world, insect pests are the major factors that cause a reduction in rice production. Unfortunately, pest problems enhanced with the intensification of irrigated rice production. Every year, around 200 million tonnes of rice are lost due to biotic and abiotic factors (Khan *et al.*, 1991). The most destructive insects of rice crop are the lepidopteran stem borers (*Tryporyza incertulas* and *T. innotata*) and the rice leaf folder (*Cnaphalocrocis medinalis*) which cause annual losses of 10 million tonnes. The occasional outbreaks of the pests can cause 60- 95% of the crop loss (Yambao *et al.*, 1993).

For many decades, insecticides have been widely used to control rice pests. However, the continuous use of a wide range of pesticides and inorganic fertilizers to intensify agriculture has resulted in many side effects, such as loss of biodiversity, the problem of secondary pest outbreaks, insecticide resistance, the resurgence of insect pests, residual toxicity and environmental pollution. Policies during the last decades have been working towards improving production methods with reduction in these negative effects. Various resource conserving strategies such as the use of compost and the integration of fish and duck production with irrigated rice are practiced for the control of weeds and insect pests (Zhang, 1992) are promoted in some countries like Malaysia (Ibrahim, 1999). Through the use of these fully or partially incompatible technologies it decreased the substantially with the introduction of pesticides. The biodiversity of irrigated rice is higher than in many natural ecosystems (Schoenly *et al.*, 1998). Many efforts have been made to combine various non-chemical control methods with insecticides. In integrated pest management (IPM) the challenge is to make natural, non-chemical controls that are effective and besides reducing some pest problems, can also be environmentally friendly. Integrated pest management practices in agriculture can be defined as an optimal combination of pest management methods that minimizes economic yield loss of a crop caused by insect pests without resulting in toxic effects to other organisms (Guo, 1998). IPM is approach that range from carefully-targeted use of chemical pesticides to biological techniques that use natural parasites and predators to control pests (Sorby *et al.*, 2003). Now a day under irrigated areas rice cropping system is facing new changes and has equally effective as the green revolution and may in turn affect the natural biological control of insect pests in rice. Predators play a significant role in the control of agriculture pests as they consume huge amount of prey and do not harm plants. However, they have not been treated as an important biological control agent, because there is so little information on the ecological role of predators in pest control.

Spiders are important predators which help to regulate the population densities of insect pests (Kajak *et al.*, 1968). Spiders have been reported to be abundant in crop fields such as rice (Tanaka, 1989), wheat (Greenstone, 2001), cotton (Ghavami *et al.*, 2008), groundnut (Munyuli *et al.*, 2008), soybean (Pearce *et al.*, 2005) and sunflower (Royer and Walgenbach, 1991), orchards (Ghavami, 2006). Furthermore, they were reported in the rose fields (Ghavami and Nematollahi, 2006), and in the surrounding landscapes of agricultural lands (Young and Lockley, 1994) as well. It has been confirmed from various research works that spiders in rice fields can play an important role as a predator in reducing the populations of insect pests of rice crop (Choi and Namkung, 1976). Research on spider diversity in agroecosystems is highly valuable; both to observe the effect of such predators have on herbivorous pests (Maloney *et al.*, 2003) and to understand how profound changes in the environment affect spider colonization (Öberg, 2007).

The spiders are carnivorous arthropods and cosmopolitan in distribution. They are classified into 40,000 species and are found all over the world in almost every kind of habitat. The population densities and species abundance of spider communities in agricultural fields can be as high as in natural ecosystems (Turnbull, 1973). A diverse group of spiders may be effective as biological control agents because they differ in hunting strategies, habitat preference and active periods. About 432 species of spiders belonging to 131 genera under 27 families were documented in the rice lands of South and Southeast Asia (Barrion and Litsinger, 1995). Young and Edwards (1990) recorded 75 species of spiders in the rice fields in the United States. Satpathi (2004) recorded 21 species of predacious spiders belonging to eight families from the seedbed to ripening of cereal crops in eastern Himalayas and found that *Lycosa pseudoannulata, Oxyopes sp., Tetragnatha* sp., *Neoscona* sp., *Argiope* sp., *Araneus inustus, Cyrtophora cicatrosa* and *Callitrichia formosana* were abundant in the rice fields. Nandakumar and Pramod (1998) reported that *Lycosa pseudoannulata, Oxyopes* sp., *Tetragnatha maxillosa, Phidippus* sp., *Callitrichia* sp., *Raneus* sp. were common in the rice fields in Kerala, India. Sebastian *et al.* (2005) recorded 92 species, 47 genera and 16 families of spiders in the irrigated rice fields in Central Kerala, India and found Araneidae and Tetragnathidae to be the dominant families with *Tetragnatha mandibulata* (Tetragnathidae) being the most abundant spider species. Sudhikumar *et al.* (2005) recorded 94 species from the rice fields of Kerala. Bharadwaj and Pawar (1987) reported *Tetragnatha mandibulata* and *Lycosa pseudoannulata* to be the important predators in a rice field in Chhattishgarh region, Madhya Pradesh, India. Rajeswaran *et al.* (2005) recorded

19 species of spiders grouped under 15 genera and belonging to 10 families in the rice fields of Coimbatore Region, Tamilnadu, India.

Their role as predators in both agricultural and natural ecosystems and potential as biological control agents which help to regulate the population densities of insect pests can only be appreciated through a greater understanding of their abundance and species composition in different ecological systems (Kaur *et al.*, 2001). In this context, the present study was done to document and quantify the spiders in rice ecosystem and study their species diversity, seasonal occurrence and abundance during the entire crop cycle.

References

Barrion, A.T. and J.A. Litsinger (1995). Riceland Spiders of South and Southeast Asia. CABI International. p. 736.

Bharadwaj, D. and Pawar, A.D. (1987). Predation of rice insect pests in Chattisgarh region, Madhya Pradesh, India. IRRN. 12: 4 p. 35.

Chiu, S.C. (1979). Biological control of the brown planthopper. In; Brown planthopper, Threat to Rice Production in Asia. International Rice Research Institute, Los Banos, Laguna, Philippines, pp. 335-355.

Choi, S.S. and Namkung, J. (1976). Survey on the spider of the Rice paddy field (1). *Kor. J. Plant Protection* **15(2)**: 89-93.

Gavarra, M. and Raros, R.S. (1973). Studies on the biology of the predator wolf spider, *Lycosa pseudoannulata* Bös. st Str. (Araneae: Lycosidae). *Philippine Entomologist* **2(6):** 427.

Ghavami, S. (2006). The role of spiders in citrus orchards in northern part of Iran. *Sonbol. J. Agri. Sci.* **150**:33-4.

Ghavami, S. and Nematollahi, M. (2006). Investigation spider fauna of rose orchards in Kashan. Proceedings of 17th Iranian Congress on Plant Protection.

Ghavami, S., Amin, G.A., Taghizadeh, M. and Karimian, Z. (2008). Investigation of abundance and determination of dominant species of spider species in Iranian cotton fields. *Pak. J. Bio. Sci.* **11(2)**: 181- 87.

Ghavami, S. (2004). The role of spiders in biological control in *Iran. Sonbol. J. Agri. Sci.* **135**: 24-5.

Ghavami, S. (2006). Investigation spider fauna of citrus orchards in northern part of Iran. Plant Pests and Diseases Research Institute. Final report of project. 36pp.

Ghavami, S., Mohammadi, D.M., Soodi, S., Javadi, S. and Ghannad, A.S. (2007). Investigation spider fauna of olive orchards in northern part of *Iran. Pak. J. Bio. Sci.* 10(15): 2562-568.

Guo, Y. (1998). Research Progress in Cotton Bollworm. Agricultural Press Beijing.

Hamamura, T. (1969). Seasonal fluctuation of spider population in paddy fields. *Acta Arachnol.* **22(2)**:40-50.

Ibrahim, M.H. (1999). Macroeconomic variables and stock prices in Malaysia: An empirical analysis. *Asian Economic Journal* **13**: 219-31.

IRRI (1989). IRRI towards 2000 and beyond. Los Banos, International Rice Research Institute Philippines, p. 68.

Kajak, A., Andrzejeuska, L., Wojcik, Z. (1968). The role of spiders in the decrease of damage caused by Acridoidea on meadows- Experimental investigation. *Ekol Pol* **16**: 755- 64.

Kaur, S., Shenhmar, M., Brar, K.S. (2001). Spider fauna of paddy in Punjab. *Insect Environment*. **7(1)**: 24-5.

Khan, Z.R., Litsinger, J.A., Barrion, A.T., Villanueva, F. D., Fernandez, N. J. and Taylo, L.D. (1991). World Bibliography of Rice Stem Borers 1794–1990. International Rice Research Institute, Los Baños, Philippines.

Lam, P. V., Huong, T. H. and Lan, T. T. (1997). A study on spider fauna of rice fields. *Nong Nghiep Cong Ngiep Thuc Pham* **3**: 107-09.

Maloney, D., Drummond, F.A. and Alford, R. (2003). Spider predation in agroecosystems: Can spiders effectively control pest populations. *Technical Bulletin Maine Agricultural and Forest Experiment Station* **190**: 28.

Miliczky, E.R. and Horton, D.R. (2005). Densities of beneficial arthropods within pear and apple orchards affected by distance from adjacent native habitat and association of natural enemies with extra-orchard host plants. *Biological Control* **33**: 249-59.

Munyuli, T.B., Kyamanywa, S. and Luther, G.C. (2008). Effects of groundnut genotypes, cropping systems and insecticides on the abundance of native arthropod predators from Uganda and Democratic Republic of Congo. *Bulletin of Insectology* **61**: 11-9.

Kumar, N. C. and Pramod, M.S. (1998). Survey of National Enemies in Rice Ecosystem. *Insect Environment* **4(1)**: 16-7.

Normiyah, R. and Chang, P.M. (1997). Pest management practices of rice farmers in the Muda and Kemubu irrigation schemes in peninsular Malaysia. In: Pest management of rice farmers in Asia, (K.L. Heong & M.M. Escalada eds.) pp. 115-27 IRRI, Los Banos, Philippines.

Oberg S (2007) Diversity of spiders after spring sowing- influence of farming system and habitat type. *J. appl. Ent.,* **13**: 524-31.

Pearce, S., Zalucki, M. P. and Hassan, E. (2005). Spider ballooning in soybean and non-crop areas of southeast Queensland Sarina. *Agriculture, Ecosystems and Environment* **105**: 273–81.

Rajeswaran, J., Duraimurugan, P. and Shanmugam, P.S. (2005). Role of spiders in agriculture and horticulture ecosystem. *J. Food Agri. Envir.* **3**: 147-52.

Royer, T.A. and Walgenbach, D.D. (1991). Predacious arthropods of cultivated sunflower in eastern South Dakota. *Journal of the Kansas Entomological Society* **64**:112–16.

Satpathi, C. R. (2004). Predacious spiders of crop pests. Capital publishing company, New Delhi. 188.

Schoenly, K.G., Justo, H. D., Barrion, A. T., Harris, M. and Bottrell, D.G. (1998). Analysis of invertebrate biodiversity in a Philippine farmer's irrigated rice field. *Environ. Entomol.* **27 (5)**: 1125-36.

Sebastian, A.P., Mathew, J.M., Pathummal, B. S., Joseph, J. and Biju, R. C. (2005). The spider fauna of the irrigated rice ecosystem in the central Kerala. *India J. Arachnol* **33**: 247–55.

Settle, W. H., Ariawan, H., Astuti, E., Cahyana, W., Hakim, A. L., Hindayana, D., Lestari, A. S. and Pajarningsih (1996). Managing tropical rice pests through conservation of generalist natural enemies and alternative prey. *Ecology* **77 (7):** 1975-88.

Sorby, K., Fleischer, G., Pehu, E. (2003). Integrated Pest Management in Development: Review of Trends and Implementation Strategies. Agriculture and Rural Development Working Paper 5. World Bank, Washington, D.C. USA.

Sudhikumar, A. V., Mathew, M. J. and Sebestian, P. A. (2005). Web building spiders of Rice Agro ecosystems in Kuttanad, Kerela. *Insect Environment* **10(3)**: 117-18.

Tanaka, K. (1989). Movement of spiders in arable land. *Plant Protection* **43(1)**: 34–9.

Threat to Rice Production in Asia. International Rice Research Institute, Los Baños, Laguna, Philippines, pp. 335-355.

Turnbull, A.L. (1973). Ecology of the true spiders (Araneomorphae). *Ann. Rev. Entomol.* **18**: 305-48.

Wu, K. M., Guo, Y. Y. (2005). The evolution of cotton pest managementpractices in china. *Annu. Rev Entomol* 50:31–52. doi: 10.1146/annurev ento.50.071803.130349.

Yambao, E. B., Ingram, K. T., Rubia, E. G. and Shepard, B. M. (1993). Case study: Growth and development of rice in response to artificial stem borer damage. *Int Crop Prot SARP Res Proc.*

Young, O. P. and Lockley, T. C. (1994). Spiders of an old field habitat in the delta of Mississippi. *Journal of Arachnology* **22**: 114–30.

Young, O. P., Edwards G. B. (1990). Spiders in United States field crops and their potential effect on crop pests. *J Arachnol* **18:**1-27.

Zhang Z Q (1992) The Use of Beneficial Birds for Biological Pest Control in China. *Biocontrol News and Information* **13 (1):** 11-6.

Chapter 2

Insect Pests of Rice

Kapila Mahajan, Amandeep Singh and Randeep Singh

Among various constraints of good rice production, infestation of different insect species is very important. The rice plant is attacked by more than 100 species of insects and 20 of them can cause serious economic loss (Pathok, 1977). Yield loss due to insect pests of rice has been estimated about 30-40% (Henrich *et al.*, 1979). The rice stem borers are the principal devastators and responsible for economic crop losses under field condition (Mahar and Hakro, 1979). They are common and serious pests in Asian countries responsible for annual damages of 5-10% of rice crops (Pathok and Khan, 1994). Heavy infestation may cause yield loss up to 80% (Rubia-Sanchez *et al.*, 1997). Five species of rice stem borers have been reported in South East-Asia namely; Dark headed stem borer (DHSB), *Chilo polychrysus* (Meyrick); Yellow stem borer (YSB), *Scirpophaga incertulas* (Walker); Pink stem borer (PSB), *Sesamia inferens* (Walker); Stripped stem borer (SSB), *Chilo supressalis* (Meyrick) and White borer (WB), *Scirpophaga innotata* (Walker) (DRR, 2006). YSB is the most destructive insect pest of rice crop (Mahar *et al.*, 1985) and responsible for an annual yield loss of 10-15% with local catastrophic outbreaks causing up to 60% damage (Catling and Islam, 1981). It is the most important insect pest of rice in Bangladesh (Islam and Hassan, 1999) and also considered as a major pest in Asia (Torri, 1971).

Although, they do not appear regularly, but cause varying degree of losses to different stages of rice in different parts of the country. On the basis of damage, insect pests are categorized into four groups.

i) Piercing and sucking insects

ii) Stem borers

iii) Leaf feeders

iv) Root feeders

Table 1. Important insect pests in rice crop

S. No.	Insect Pest	Type of damage
		A. Piercing and sucking pests
1.	**Brown plant hopper**[1] *Nilaparvata lugens*[2] **Hemiptera**[3] **Delphicidae**[4]	Both the adults and nymphs cause damage by sucking cell sap from the leaves which turn yellow. If the insect attacks during early stages of growth, entire plant may dry up. In the later stages, the tillering is adversely affected, heavy infestation causes drying of the crop in patches called "hopper burn". This insect act as vector of that transmits grassy stunt virus.
2.	**Whitebacked planthoppers**[1] *Sogatella furcifera*[2] **Hemiptera**[3] **Dephacidae**[4]	The nymphs and adults suck cell sap, particularly from the leafsheath. The leaves of the attacked plants turn yellow and later on rust red; symptoms starting from leaf tips move downwards. The attacked plants ultimately dry up in patches without producing ears.
3.	**Green leafhopper**[1] *Nephotettix nigropictus*[2] **and** ***N. virescencs***[2] **Hemiptera**[3] **Cicadellidae**[4]	The nymphs and adults suck the cell sap from the leaves. The plants lose vigour, turn yellow and ultimately brown. *N. virescens* is also a vector of tungro virus.
4.	**White rice leafhopper**[1] *Cofana spectra*[2] **Hemiptera**[3] **Cicadellidae**[4]	Both adult and nymphs suck cell sap and cause yellow discoloration of the leaves. When infestation is heavy, leaves turn brown, plant growth stunted and fail to produce ears.
5.	**Zig-zag leafhopper**[1] *Recilia dorsalis*[2] **Hemiptera**[3] **Cicadellidae**[4]	As a result of feeding on cell sap by insect pest, mature leaves acquire an orange discolouration of the margins and become dry at the tips.
6.	**Rice blue leafhopper**[1] *Typhlocyba maculifrons*[2] **Hemiptera**[3] **Cicadellidae**[4]	Both nymphs and adults suck cell sap from the leaves which, in the early stages of an attack, cause whitish, waxy lines. As the damage progress, the leaves show symptoms of withering.
7.	**Rice bug**[1] *Leptocorisa acuta*[2] **Hemiptera**[3] **Coreidae**[4]	The nymphs and adults suck juice from the developing grains in the milky stage, causing incompletely filled panicles with empty grains. Black or brown spots may appear around the holes made by bugs on which a sooty mould may develop. The severely infested fields emit an offensive smell
8.	**Rice mealy bug**[1] *Heterococcus rehi*[2] **Hemiptera**[3] **Coccidae**[4]	Both the nymphs and adult suck plant sap from the stem, resulting in stunted plant growth and yellowish curling of leaves. When the attack is severe, the ear heads become smothered and are unable to grow out of their sheath.
9.	**Rice thrip**[1] *Stenchaethrips biformis*[2] **Thysanoptera**[3] **Thripidae**[4]	The larvae and adults lacerate the tender leaves and suck the plant sap, causing rolling and drying of the leaf tips

Note- 1- Common Name, 2- Scientific Name, 3- Order and 4- Family

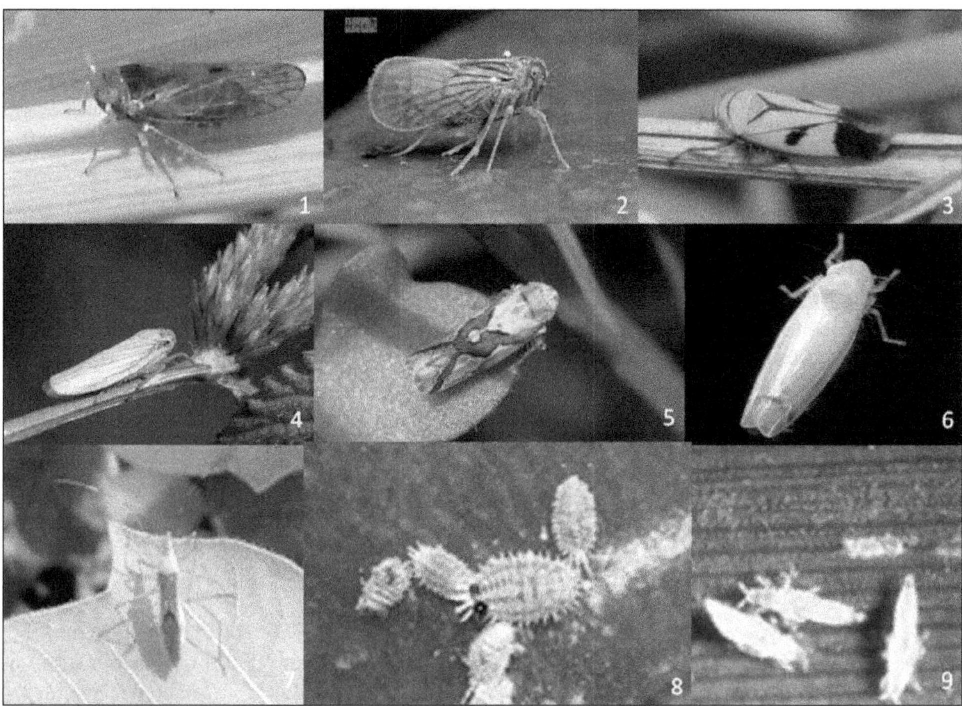

Plate-1. Sources: 1. (www.pestnet.org), 2. (Hafiz M 2012), 3. (www.cabbsouat.org), 4. (www.inaturalist.org), 5. (http://blog.naver.com), 6. (Ashley Bradford, 2012), 7, 8 & 9. (www.cabbsouat.org)

	B. Stem borers	
10.	**Yellow stem borer**[1] ***Scirpophaga incertulas***[2] **Lepidoptera**[3] **Pyralidae**[4]	The larvae feed inside the stem, causing drying of the central shoot or dead heart in the young plants and drying of the panicle or white ear in older plants
11.	**Pale headed striped borer**[1] ***Chilo suppressalis***[2] **Lepidoptera**[3] **Pyralidae**[4]	The injury to the plant is caused by caterpillars which tunnel through the stem and feed on the soft tissues, causing dead hearts in early stages and white ears in later stages.
12.	**Dark headed striped borer**[1] ***Chilo polychrysus***[2] **Lepidoptera**[3] **Pyralidae**[4]	The caterpillars bore into the central shoot for feeding. Since the larvae bore into the central shoot for feeding. The larvae bore into the outer leaves and leafsheaths first, these are first to die, followed by inner whorl, and finally the entire plant from its core
13.	**Pink borer**[1] ***Sesamia inferens***[2] **Lepidoptera**[3] **Noctuidae**[4]	The damage is caused by caterpillars and attacked young plants show dead hearts and are killed altogether.

Note- 1-Common Name, 2- Scientific Name, 3- Order and 4- Family

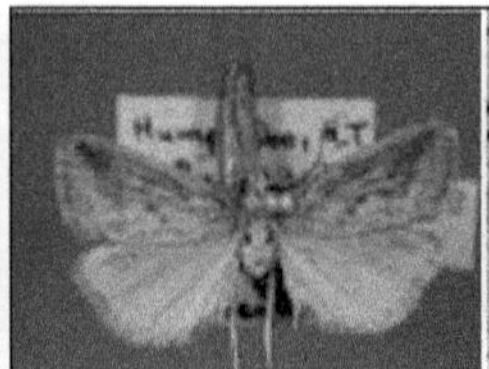
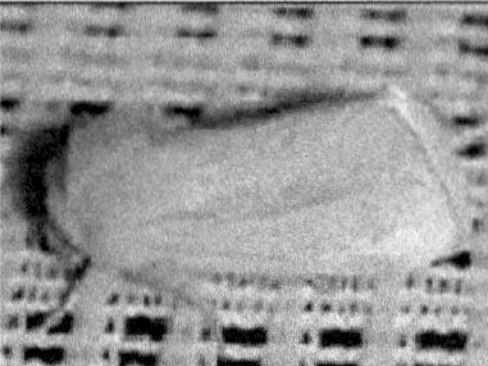

Plate-2. Source: 10, 11, 12 & 13. (www.cabbsouat.org)

	C. Leaf Feeders	
14.	**Rice leaf folder**[1] *Cnaphalocrocis medinalis*[2] **Lepidoptera**[3] **Pyralidae**[4]	The young larvae feed on tender leaves without folding them. The older larvae fasten the longitudinal margins of leaf together with a sticky substance and feed inside the fold by scraping the green matter. The scraped leaves become membranous, turn white and finally wither.
15.	**Rice ear cutting caterpillar**[1] *Mythimna separata*[2] **Lepidoptera**[3] **Noctuidae**[4]	The newly hatched larvae feed on the epidermis of the tender leaves. The second and third instar larvae feed by cutting the leaf from the edge towards midrib. The older larvae, besides damaging leaves, also cut off the panicles, mostly at the base.
16.	**Rice caseworm**[1] *Nymphula depunctalis*[2] **Lepidoptera**[3] **Pyralidae**[4]	The insect attacks the crop in the early transplanted stage. The leaf blade is cut into small bits and a tubular case is constructed by the larva. The larvae feeds by scraping the undersurface of the leaf blade leaving the upper epidermis intact, resulting in white patches on leaf blades.
17.	**Paddy gall fly**[1] *Orseolia oryza*[2] **Diptera**[3] **Cecidomyiidae**[4]	The damage is caused by maggots which feed inside the stem. Besides causing physical damage, they produce galls on the basal portion of the central leaf or on the tillers. These galls ultimately become hollow giving a silvery shine called silver shoots. The infested tillers do not bear tillers.
18.	**Rice hispa**[1] *Dicladispa armigera*[2] **Coleoptera**[3] **Chrysomelidae**[4]	The grubs mine into the leaves presenting blister spots towards leaf tip. The adults feed by scraping the green matter and produce parallel whitish streaks on the leaves. The damage starts in nurseries and spreads to the rice fields.
19.	**Rice grasshoppers**[1] *Hieroglyphus banian*[2] **and** *H. nigrorepletus*[2] **Orthoptera**[3] **Acrididae**[4]	The adults and nymphs feed on the leaves, leaving midribs and stalks. At the shot blade stage of the crop, they nibble at the florets or gnaw into the base of inflorescence stalks causing formation of white ears.
20.	**Whorl maggot**[1] *Hyrellia philippina*[2] **Diptera**[3] **Ephydridae**[4]	The maggots attack the leaf blades even before unfurling and initial damage is characterized by the presence of narrow stripes of whitish area in the blade margins. The tillers become stunted.

		D. Root Feeder
21.	**Rice root weevil**[1] ***Echinocnemus oryzae***[2] **Coleoptera**[3] **Curculionidae**[4]	The grubs of the rice root weevil feed on the root hairs of the transplanted paddy crop. The plants become weak, stunted, pale yellow and ultimately dry up. The removal and destruction of stubble helps to reduce the hibernating insects.

Note- 1-Common Name, 2- Scientific Name, 3- Order and 4- Family

Plate-3. 14-16. (www.cabbsouat.org), 17. (www.knowledgebank.irri.org), 18-20. (www.cabbsouat.org), 21. (pbt.padil.gov.au)

Pest management in rice

Brown plant hopper considered as a minor pest until green revolution, but after the green revolution during the seventies, this pest became a major pest in rice (Sigsgaard, 2000). Matteson (2000) reported the effects of "turning off" the biological control of this pest, which is maintained at low levels by spiders and other natural enemies. When spiders and other predators were removed, the population of BPH increased drastically (Kenmore *et al.*, 1984). The intensification of insecticides resulted in the emergence of insecticide resistant strain of insect pest. When resistant rice varieties against BPH and other pests were planted over large areas, then these pests overcome the plant resistance (Heinrichs and Mochida, 1984). Subsequently, a new pest management strategies such as Integrated Pest Management (IPM) were developed that emphasized host plant resistance, biological control and minimal use of insecticides (Waage, 1999). An increasing amount of research evidence from tropical irrigated rice areas shows that there is little or no crop loss in insecticide untreated fields (Kenmore, 1991; Litsinger, 1991; Way and Heong, 1994). This inconsistency may be explained by: a) earlier estimates of yield loss were based more on damage than on actual yield, b) moderate resistance against insect pests in many modern varieties of rice, c) the ability of some modern varieties to compensate for damage, because they produce more tillers (Rubia *et al.*, 1989), d) better control of insect pests by natural enemies with less use of insecticides (Way and Heong, 1994). Previous research showed that moderately BPH-resistant and BPH-susceptible rice varieties grown by a large number of farmers have had low and stable BPH populations for several years suggest that the pest control strategy in rice should be revised to put higher priority on natural biological control (Heong and Schoenly, 1998). Apart from the fact that use of insecticide is rarely necessary, it also poses a risk to farmer health and the environment (Heong *et al.*, 1995). Continued use of insecticide stresses the need to bridge the gap between research and farmers. In many countries, FAO (Food and Agriculture Organization) has supported Farmers' Field Schools and provided farmers with a practical understanding of integrated pest and nutrient management (Matteson, 2000). The expectation of this program was that the farmers who receive training will pass their new knowledge on to other farmers. Another approach was developed by Heong *et al.*, (1998) in which farmers were motivated to 'test' a simple rule of thumb (no spray necessary in the first 40 days after sowing) by the use of communication media, including the radio. The practice of no early spray is now adopted by many farmers in southern Vietnam and recommended by the National Agricultural Research and Extension Agencies in Malaysia, The Philippines and Thailand. The rice fields are robust and stable even in the absence of insecticides as a result of an extremely rich web of generalist natural enemies, mainly predatory spiders (Premila, 2003). As generalists, spiders are ingenious predators and their inherent ability to regulate pest numbers is ineffable and thus heaves the role of these carnivorous arthropods in rice pest management.

Natural enemies

Like any other animals, spiders have their own foes. Among these are 15 hymenopteran natural enemies having six undescribed species of scelionids belonging to three genera *(Baeus, Ceratobaeus* and *Idris),* six ichneumonid wasps in five genera *(Astomaspis, Caenopimpla, Linella, Paraphylax* and *Strepsimailus),* two sphecid wasps, *Chalybion bengalense* (Dahl born) and *Sceliphron madraspatanum conspicillatum* Costa and two undetermined braconid genera, one Pompilinae and one each of a mantispid (Neuroptera: Mantispidae) and a sarcophagid fly (Diptera: Sarcophagidae). The small scelionid wasps are the most common egg parasitoids of the leaf folding spider, *Clubiona japonicola* Boesenberg and Strand and the orb-weavers, *Araneus inustus* (L. Koch) and *Neoscona theisi* (Walckenaer). The ichneumonid *Caenopimpla arealis* (Cushman) was specific to the lynx spider, *Oxyopes javanus* Thorell whereas the other five ichneumonids were reared from *Argiope catenulata, Atypenaformosana* and *Tetragnatha* spp. So far the braconids were only reared from *Tetragnatha.* The undetermined mantispid was reared for the first time from the egg cocoons of *P. pseudoannulata.* A sarcophagid fly, *Pierretia litsingeri* Shinonaga and Barrion was host specific to *A. catenulata.* Pathogens and nematodes also parasitize spiders. The fungus *Gibellula leiopus* (Vuill.) Mains, *Torrubiella* sp. and one undetermined species attacked spiders. All these fungi were observed on *A. formosana, Clubiona* and salticid spiders. On the other hand, the nematode was isolated from the Iycosid, *Pardosa* spp.; salticid, *Cosmophasis* sp.; and the thomisid, *Thomisus okinawensis* Strand. Based on field observations and rearings of spiderlings and adults, 0.6-56% level of parasitism had been recorded. The other predators of spiders are bull frogs, toads, birds, ants. Similarly, the marsh fly, *Sepedon* spp. preyed on the eggs of *Tetragnatha* spp. Some spiders are araneophagic hunters, so they themselves are their predators. Natural enemies are strong limiting factors in the effective and efficient utilization of spiders in natural biological control of rice insect pests.

References

Catling, H.D. and Islam, Z. (1981). Problem of yellow stem borer in Asam deep water rice. *Proceeding International Rice Research Institute* 451-58. .

Heinrichs, E.A. and Mochida, O, (1984), From secondary to major pest status: the case of the insecticide induced rice brown planthopper, *Nilaparvata lugens,* resurgence *Protection Ecology* **7:** 201-18.

Heinrichs, E. A., Sexena, R.C. and Chelliah, S. (1979). Development and implementation of insect pests management systems for rice in tropical Asia. ASPAC Bulletin 127. Taiwan: Food and fertilizer technology center.

Heong K L and Schoenly K G (1998) Impact of insecticides on herbivore-natural enemy communities in tropical rice ecosystems. In: Ecotoxicology: pesticides and beneficial organisms, (P.T. Haskell & P. McEwen eds.), pp. 381-403. Kluwer, Dordrecht, Netherlands.

Heong, K. L., Escalada, M. M. and Lazaro, A. A. (1995). Misuse of pesticides among rice farmers in Leyte, Philippines. In: Impact of pesticides and farmer health and the rice environment (P.L. Pingali & P.A. Roger eds.), pp. 97-108. IRRI, Los Banos, Laguna.

Islam, Z. and Hasan, M. (1999). Pests of rice in Bangladesh: Present management scenario and future challenges. pp. 90-98, Proc. First Agric. Cont. CARE Bangladesh Dhaka.

Kenmore, P. E. (1991). Indonesia's integrated pest management – a model for Asia. Food and Agriculture Organisation, Manila, Philippines.

Kenmore, E., Carino, F., Perez, C., Dyck, V. and Gutierrez, A. (1984). Population regulation of the rice brown planthopper (Nilaparvata lugens Stal) within rice fields in the Philippines *Journal of Plant Protection in the Tropics* **1:** 1-37.

Litsinger, J. A. (1991). Crop loss assessment in rice. In: Rice insects: management strategies (E. A. Heinrichs & T.A. Miller eds.), pp. 1-65. Springer Verlag, New York.

Mahar, M. N., BrattiI, M. and Dhuyo, A. R. (1985). Stem borer infestation and yield loss relationship in rice and cost- benefit of control. Paper presented at 5th National Seminar on rice production.

Mahar, M. M. and Hakro, M. R. (1979). The prospects and possibilities of yellow rice stem borer eradication under Sindh condition. Paper presented at the Rice Research and Production Sem. pp 18- 22, Islamabad.

Matteson, P.C. (2000). Insect pest management in tropical Asian irrigated rice *Annual Review of Entomology* **45**: 549–74.

Pathok, M. D. (1977). Defense of the rice against insect pests. *Ann. N.Y. Acad. Sci.* 287-95.

Pathok, M. D. and Khan, Z. R. (1994). Insect pests of rice. International Rice Research Institute, Los Banos, Philippines.

Premila, K. S. (2003). Major Predators in rice ecosystems and their potential in rice pest management. Ph. D. thesis, Kerala Agricultural University, Thrissur, **43:** 57- 72.

Rubia, E. G., Shepard, B. M., Yamba, E. B., Ingram, K. T., Arida G S, Penning de Vries F (1989) Stem borer damage and grain yield of flooded rice *Journal of Plant Protection in the Tropics* **6:** 205-11.

Rubia-Sanchez, E. G., Nurhasyim, D., Heong, K. L., Zaluki, M., Norton, G. A. (1997). What stem borer damage and grain yield in irrigated rice in west Java, Indonesia. *Crop protection* **16:** 665-71.

Sigsgaard, L. (2000). Early season natural biological control of insect pests in rice by spiders - and some factors in the management of the cropping system that may affect this control *European Arachnology* **4:** 57–64

Torii, T. (1971). The ecological studies of rice stem borers in Japan: a new review Mushish **45**: 1-49.

Waage, J. (1999). Beyond the realm of conventional biological control: harnessing bioresources and developing biologically-based technologies for sustainable pest management. In: Biological control in the Tropics. Towards efficient biodiversity and bioresource management for effective biological control. Proceedings of the Symposium on Biological Control in the Tropics, pp. 5-17. The National Council for Biological Control (NCBC) /CAB International, South East Asian Regional Centre, Malaysia.

Way, M. J. and Heong, K. L. (1994). The role of biodiversity in the dynamics and management of insect pests of tropical irrigated rice: A review. *Bulletin of Entomological Research* **84:** 567-87.

Chapter 3

Brief about Spiders

Amandeep Singh and Randeep Singh

Classification

Like the scorpions, the true spider (Araneae) have a place with the class Arachnida (arachnids). They are the biggest order in this class, with around 30,000 species, and are partitioned into the 3 suborders Mesothelae, Mygalomorphae (Orthognatha) and Araneomorphae (Labidognatha), and further into various families. The Mesothelae are a little group of Southeast Asian spiders that have scarcely any clinical significance. Some risky species are found among the Mygalomorphae (Bird insects in the more extensive sense), however it is Araneomorphae (Labidognatha), by a far the biggest suborder, which incorporates the greatest number of hazardous species.

Common Name:	Spiders
Kingdom:	Animalia
Phylum:	Arthropoda
Class:	Arachnida
Order:	Aranea
Family:	101 families
Genus species:	40,000 species (approximation)

(source: https://seaworld.org/Animal-Info/Animal-Bytes/Arthropods/Spiders)

(source: http://www.vapaguide.info/page/22)

Morphology of Spider

Spiders have two fundamental body parts (appeared in plate-4.): the cephalothorax, which may likewise be alluded to as the prosoma (1), and the abdomen, which may likewise be referred to as the opisthosoma (2). Protruding from the backside of the abdomen are the spinnerets (3), while distending from the front of the cephalothorax are the chelicerae (4). The arachnid eyes (5) are situated on the cephalothorax. Although the vast majority realize that spiders have eight legs, numerous are confounded when they see a live spider, since it might seem to have ten legs. This additional combine of "legs" is really not legs, but rather pedipalps (6). The "genuine" legs are the four sets of walking legs, which are numbered 1(7), 2(8), 3(9), and 4(10) (https://academics.skidmore.edu/wikis/NorthWoods/index.php/Spider_morphology).

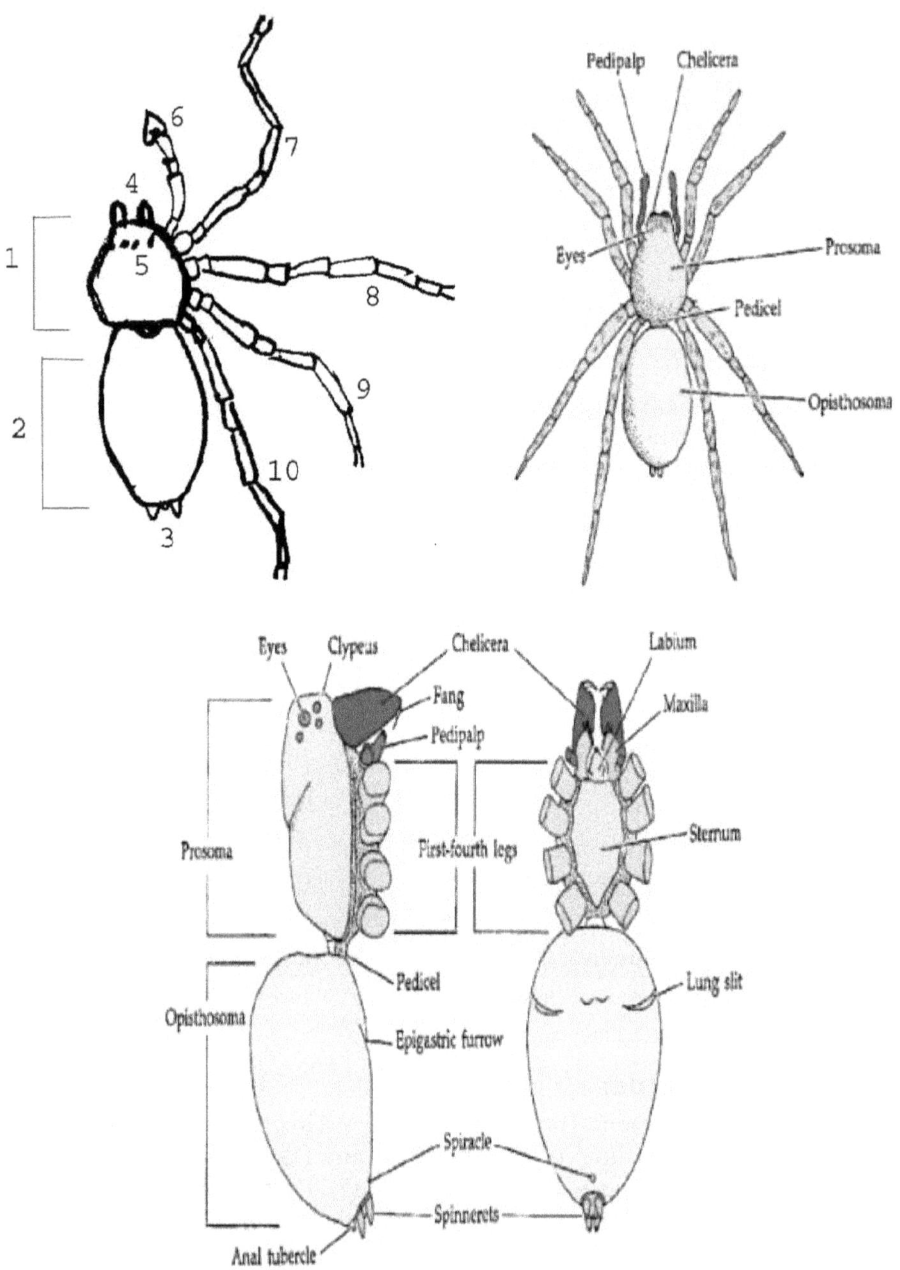

Plate-4: Morphology of Spider

The Cephalothorax

The forepart of the spider's body named as prosoma and furthermore known as cepholothorax. The cephalothorax (plate-5.) is a nearly uniform' structure. The shield or carapace which limits it above is once in a while a smooth, general arched surface, however more regularly a visible groove partitions a clear head from the thorax behind it. Upon this thoracic area there are for the most part spaces—a "middle fovea" and eight "radial striae" pointing towards the legs. These depressions mark the inside connections of the muscles of the sucking stomach and of the legs; they are regularly more profound in colour than the encompassing shield and may frame the only pattern borne by the cephalothorax. Now and then, be that as it may, dark longitudinal streaks are present; indeed, in numerous families there is a more or less standard pattern to which its individuals accommodate. Again, in a few species the cephalothorax is encompassed by spines (Savory, 1928).

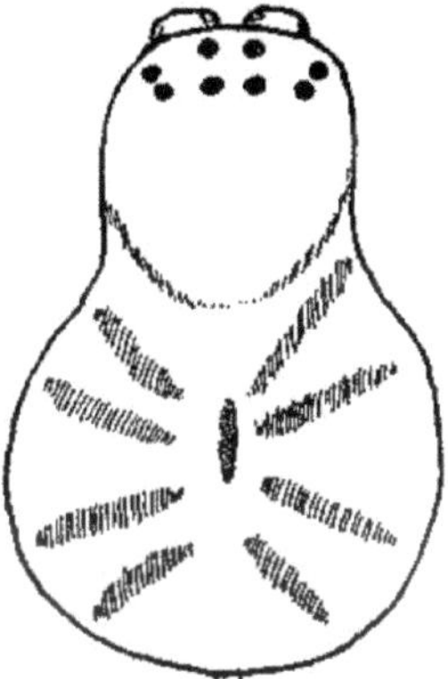

Plate-5: Spider's Cephalothorax

The Pedicle

The cephalothorax and prosoma area are joined by the typically slender waist or pedicle, covered as a rule by the overhanging-abdomen. This sensitive junction is ensured and strengthened by chitinous plates above and underneath, known as the lorum and plagula respectively. The shape of both lorum and plagula are various, but there seems to be no principle governing which are in sight in various families. The lorum is frequently made out of two pieces, which fit nearly to each other, yet the plagula is constantly unified.The cephalothorax and prosoma are joined by the naturally slender waist or pedicle, hidden as a rule (Savory, 1928).

The Abdomen

The ordinary abdomen (Plate-6.) area or opisthosoma is a more or less prolonged tube shaped sac, devoid of all traces of segmentation and very often with no pattern. The greatest conceivable decent variety is, be that as it may, found. Pattern and frequently beauty of colouring and configuration are obvious in numerous families, and where a pattern or marking of any kind exists, three general features may usually be recognised.

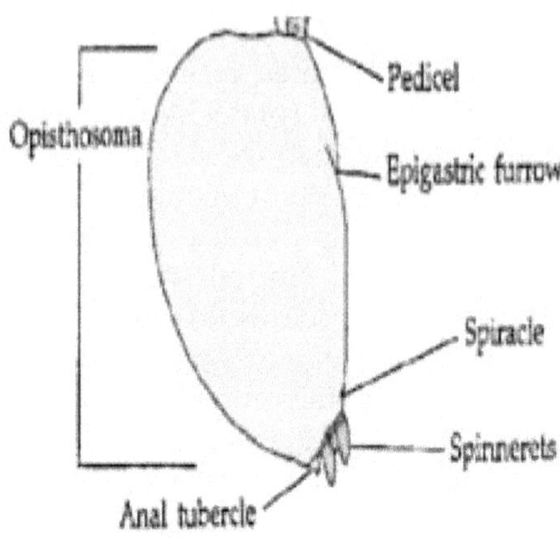

Plate-6: Abdomen of Spider

Most as often as possible, a longitudinal tight dorsal mark is present, lying over the heart inside and maybe because of its proximity. In different families, particularly the orb-spinners, a more extensive leaf-formed mark is found, and is known as the folium. Thirdly, little discouraged points, solidified inside, and due, similar to the striations of the cephalothorax, to inward muscle connections, are regularly obvious symmetrically organized, and are seen most effectively on spiders without different markings (Savory, 1928).

The Sternum

The underside of the prosoma is shaped by two unequal plates of chitin named the sternum and labium or hp (Plate-7), the previous is oval or heart-formed, somewhat curved and as a rule marked on each side by four shallow bays or acetabula, opposite the coxae of the legs. Like the carapace, the sternum represents to various fused segmental plates (Savory, 1928).

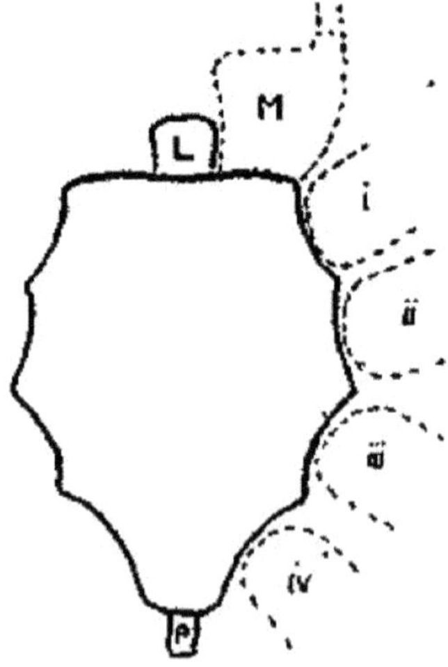

Plate-7: Spider's sternum

The Chelicerae

The appendages of the cephalothorax are the chelicerae, the palpi and the legs. The chelicerae (chelae, mandibles or falces) are the spiders' extremely effective weapons (Plate-8). Here it might be noticed that the quantity of alternative names for relatively every organ is a normal for illustrative anatomy in spiders. The chelicerae are homologous with the second antennae of crustacean and not with the mandibles of insects. They comprise constantly of two joints, the proximal one being named the paturon or tige (Savory, 1928).

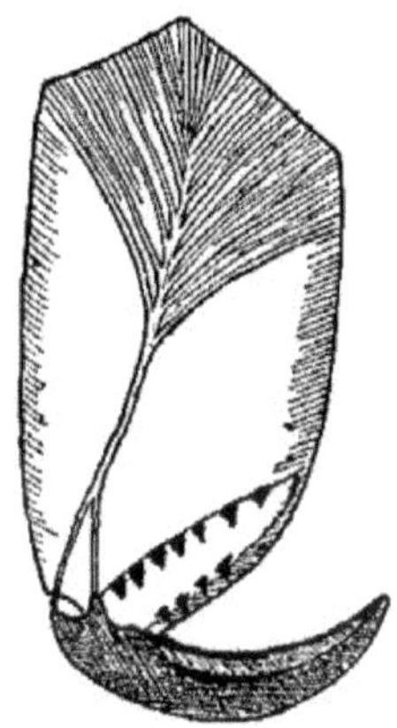

Plate-8: Spider's chelicerae

The Palpi

The second pair of appendages, the palpi, performs assorted functions. They are six-jointed appendages, the joints being coxa, trochanter, femur, patella, tibia, and tarsus. In 'the Spiders of Dorset, Pickard-Cambridge does not independently name the coxa, regarding it as a part of the maxilla and names the staying five axillary, humeral, cubital, radial and digital. The distal joints are utilized in stridulation and the tarsus is an accessory to the reproductive system. Moreover there is, with the exception of in the Mygalomorphae, an endite or inside lobe of the coxa which acts as one of the mouth-parts. This endite, the maxilla or maxillary lobe, is isolated by membrane from the coxa of the appendage. Its function is to compress the nourishment particles and crush out their fluid substance into the pharynx. Its innermost margin is for the most part furnished with hairs which might be adequately thick to shape a scopula (plate-9) and the fore-edge frequently bears a serrula or line of teeth which, without a doubt, help in cutting the food (Savory, 1928).

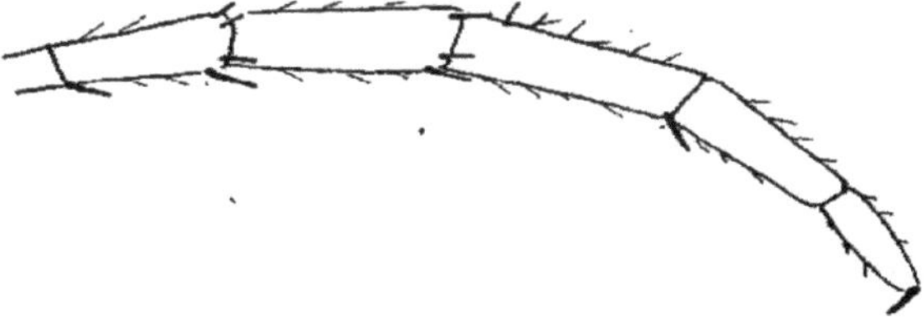

Plate-9: Spider's palpi

The Legs

The legs, constantly eight in number, are seven-jointed and/the joints are coxa, trochanter, femur, patella, tibia, meta-tarsus and tarsus (Plate-10). It is intriguing to notice that while these are the old names acquired from vertebrate morphology, the tarsus has been placed beyond the metatarsus (Savory, 1928).

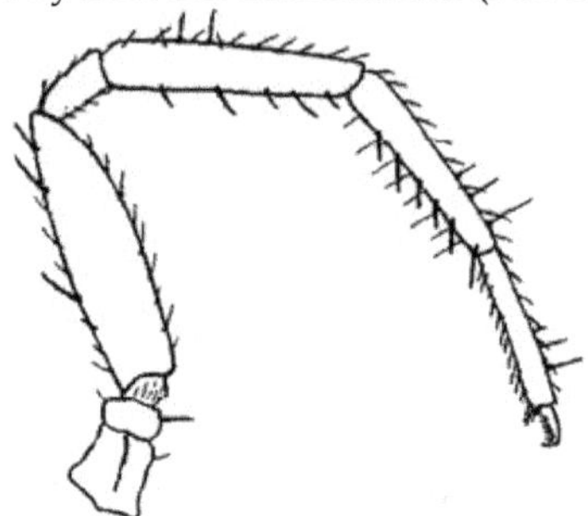

Plate-10: Spider's legs

The Setae

A significant part of the spider's body and in addition its legs, is secured with what might customarily be called hairs (plate-11). Hair in the genuine zoological sense of a living outgrowth from the skin is, in any case, particular to warm blooded animals, and the comparative belonging of the spider are better named setae. Presumably all the setae on an spider are pretty much created as sense organs, yet a portion of those on the legs are helpful assistants to the spinning organs. On looking at a spider it is anything but difficult to recognize hairs of no less than three various types. The most obvious are the stout sharp spines on the legs and palpi, by and large portrayed as tactile. The most hard to recognize, even under the magnifying instrument, are the long sensitive acoustic setae believed to be receptors of sound waves.

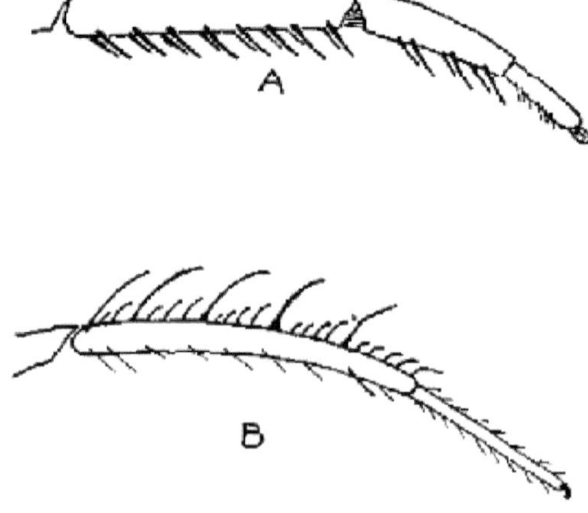

Plate-11: Arrangement of legs spines, A. *Zora spinimana*, B. *Ero furcata*

Anatomy of Spiders

The alimentary channel, to which reference has been made, is in spiders a muddled framework charged with imbibing, storing, and digesting the sustenance (Fig-6). It is of a type peculiar to Arachnida, and does not resemble that of some other class of invertebrates. The mouth is an extremely little aperature, hard to recognize clearly. Lying directly over the labium, and in close contact with it, is a flattened cone of tissue called the rostrum. On the off chance that the rostrum and labium are isolated the lower surface of the previous apparently is secured with a chitinous plate, the epipharynx. Opposed to it, on the upper surface of the labium is a comparing plate, the hypopharynx. The epipharynx and, in Mygalomorphae, the hypopharynx, are marked with fine groove forming, when put against each other, the stomodaeum up which the sustenance rises into the oesophagus, partly by surface tension, mostly by the sucking activity of the stomach inside. The epipharynx is also marked with fine transverse striations and edged with minute teeth. The alimentary channel of spiders agrees, nonetheless, with that of most other Arthropoda in being divisible into three regions, of which just the middle part is fixed with epithelium and is absorptive in real life (Savory, 1928).

The fore and hind parts are derived from the invaginations of the exoskeleton at the front and back end of the embryo, shaping the stomodaeum and proctodaeum, respectively. The fore-gut, or stomodaeum, comprises of pharynx, oesophagus, and sucking stomach. Every one of these parts is lined with chitin and the structure of their sides is the same as that of the body-wall, with which they are in continuity. The pharynx rises vertically between the epipharynx and the hypopharynx. The curvature of the epipharynx produces a space which is involved by a median gland, the pharyngeal gland. This is a minute oval mass of secretory cells with a duct leading to end of the pharynx, close to its junction with the oesophagus. The oesophagus is a significantly easier part of the canal to obtain in a dissection. It is a somewhat curved tube, whose interior chitin is thickened above and striated on the sides typically. Its lower surface is more slender and less conspicuous so the part is frequently contrasted with an inverted gutter. The sucking stomach is incorrectly named since it has none of the functions of a stomach, yet is fairly to be contrasted with a pump drawing in the food from outside. Consequently usually alluded to as the "so-called stomach," yet this appears a melticulous utilization and has not been followed here. It is shaped by an enlarging of the oesophagus, and lies on the endosternite. Almost the entire of its upper surface is hardened, forming a leaf-shaped shield with its point advances and with a middle edge beneath, so its crosssection is T-formed. This ridge is a continuation of the dorsal thickening of the oesophagus. The sucking organ is extended by the opposite muscles attached to its shield and to the midian groove of the carapace. It is shut by a progression of semicircular compressor muscles attached to the edges of the shield and to the endosternite. There is no muscle in its own composition. The mid-gut, or mesenteron, is the genuine absorptive region, and it is in this part that the spiders demonstrate their most striking departures from the more general

Arthropodan type of an unbranched tubular canal. To make nourishment valuable to the spider, the procedure of digestion must include such substance changes that will make the substance dissolvable, thus ready to go in through the intestinal walls into the blood which will distribute it to the tissues (Savory, 1928).

In different creatures, there are different devices by which the absorption of the food-products is rendered as complete as possible, either by increasing the time spent in the engrossing region or by expanding the retaining surface area in contact with the food. The former type is illustrated by the dogfish, whose comparatively short in digestive system encases a winding valve. In going down the turns of this winding the food takes a very much longer time than it would in passing directly from end to end. The second method is illustrated by the earthworm, whose intestine possesses a dorsal infolding or typhlosole. This to a great extent expands the absorptive area in contact with the finely divided soil which is passing down it. The two methods are combined in warm blooded creatures and many others, whose small intestine or ileum is greatly elongated, until its total length becomes many times greater than that of the creatures itself. The spider possesses two such devices for securing an increased efficiency, which are unlike those of any other type of creature. The alimentary canal leaves the sucking stomach as a narrow tube directed towards the pedicle. Before reaching the pedicle, there arise from its sides two diverticula or indiscriminately finishing tubes which run forwards above the endosternite as far as the poison glands (Savory, 1928).

In a few spiders, these diverticula meet in front shaping a complete circle, however generally speaking their ends, however lying near one another, are isolated. What's more, four short lateral caeca emerge from the external side of every diverticulum in the ways of the legs. They might be drawn out some little path into the coxae, or they might be bent downwards and inwards under the ventral nerve mass which lies underneath the oesophagus (Plate-12). The fluid substance of these caeca has a digestive activity on meat. They in this manner most likely go about as a store for this liquid. The mesenteron at that point goes back through the pedicle and shortly subsequent to entering the abdomen curves upwards and broadens. In the upper surface of this wider portion are generally four openings driving into the complex system of branched tubules which shape the abdominal gland. This gland involves most of within the spider's abdomen, about the entire of the upper and sidelong portions, It is entered every which way by the Malpighian tubes, and its branches ramify round the heart and digestive tract in a baffling disarray. It has since quite a while ago pulled in the considerations and hypotheses of anatomists and has been progressively portrayed as a stomach, a fat-body, a liver, and a pancreas. Reality would appear to be that it functions in two unmistakable ways. It goes about as a digestive gland, emitting an ferment upon the nourishment, yet additionally as a store, for the sustenance items go into the tubes themselves. In this way they swell out, and a spider after a large meal becoming enlarged to a degree which would be very outlandish on the off chance that it were expected just to the development of the mid-gut itself. It is, obviously, not exceptionally normal to find the nourishment entering the digestive glands rather than simply accepting their emissions through a duct, and the outcome is that it concedes the spider capacity to get moderately tremendous amounts of sustenance at once. This is stored and continuously consumed by the abdominal gland, with the goal that extensive stretches of fasting can be survived. The mesenteron goes into the proctodaeum with no incredible change in estimate, however the last mentioned, notwithstanding its chitinous coating, is encompassed too by a layer of muscle cells. It bears on its dorsal surface an enlargement, or empty, the stercoral pocket, where fecal issue gathers as a smooth liquid in which coast little dark particles (Savory, 1928).

The rectum is a straight tube opening at the anus, which lies toward the finish of a little tubercle, behind the back combine of spinnerets. The digestion of the nourishment is of need took after by three outcomes. A little extent of the nutriment picked up is stored as fat, the rest must be passed on to every one of the tissues by the blood, and the waste matter must be disposed of. The fat tissue of spiders comprises of cells containing droplets of fat. These are found in three circumstances. A layer of fat-cells lines the inside of the cephalothoracic caeca, and the space in the abdomen between the branchings of the abdominal gland is loaded with fatty material (Savory, 1928).

The Vascular System

In mammals, for example, ourselves, the normal for the circulatory framework is that there are two independent circulation streams through the heart, one of purged blood heading off to the tissues of the body, and one of blood which, after its arrival from the circuit of the body, is setting off to the lungs to be oxygenated. In spiders, as in other Arthropoda, there is nevertheless one course of blood through the heart, and one circuit of the body. This circuit, as well, is inadequate. The blood isn't at all focuses kept to vessels: there are no vessels and the interior organs lie washed in blood. The heart (Plate-12) is a straight tube, conelike fit as a fiddle, lying in the dorsal part of the abdomen, some of the time very, near the skin and some of the time implanted in the alimentary caeca. It is mostly made out of muscle cells, the majority of which are transversely organized, yet a couple are longitudinal. Outside the muscle is a layer of connective tissue strands. The heart is basic inside, not partitioned by valves into chambers. It lies in a thin-walled sac, the pericardium. Which encompasses it at some little separation in order to leave a pericardial space between the two. Both heart and pericardium are held in position by a complex arrangement of various ligaments, above, below, and at the sides. The pericardial space is loaded with blood which enters the heart through three pairs of apertures or ostia four pairs in Mygalomorphae. These ostia are given vth valves which prevent the blood from re-entering the pericardium from the heart, with the goal that it is compelled to go out of the heart by the arteries which lead from it. The aorta, or forward prolongation of the heart, plunges down and goes through the pedicle into the cephalothorax. Here the back dorsal arteries emerge from it, to supply the muscles of that region. Behind the stomach it partitions into two branches which lie between the sides of the stomach and the cephalothoracic caeca (Savory, 1928).

Close to the front end of the endostemite there emerge two forwardly coordinated cephalic veins, which supply blood to the eyes and poisonglands, while the Aortic vessels plunge abruptly downwards and form a centre from which veins raced to the palpi and legs. The lateral arteries emerging from the heart are eight in number in Mygalomorphae and six in most different spiders. They circulate blood among most of the organs contained in the abdomen. Posteriorly the heart is proceeded into the caudal course. This branches among the spinnerets and silk organs. The blood does not return by veins. It is gathered in rather vague channels called lacunae, Mrhich convey it to spaces called sinuses. There are six of these sinuses, three in every division of the body. The three in the cephalothorax are longitudinal spaces lying parallel to each other, near the sternum. Two of the abdominal sinuses are likewise close to the ventral surface, the third is beneath the pericardium. All these sinuses lead the blood to the lungbooks, where it is re-oxygenated by the air entering through the clears out. By two pulmonary veins, or, in Mygalomorphae, by four pulmonary veins, the blood currently flows back to the pericardium, whence it re-enters the heart by the ostia. These pulmonary veins are the only vessels in the spider to be called veins. They are comparative in constitution to the cardiovascular ligaments which hold the heart in place. Causard has to be sure proposed that the other lateral ligaments are reduced veins, which have lost their unique function of passing on blood and become mere ligaments (Savory, 1928).

The Reproductive System

The inside reproductive organs of spiders are not extremely complex. The testicles of the male lie parallel to each other in the abdomen beneath the alimentary canal. They are tubular in frame, closed behind, and proceeded in front into a couple of much coiled tubes, the vasa deferentia. These join at their furthest points to frame a short vesicula seminalis, leading specifically to the single median orifice in the epigastric furrow. The ovaries involve a relating position, however they are significantly bigger, particularly when the eggs are about develop, and accordingly they are much easier to find. As a result of the ovarian follicles, which project from their surfaces, they are constantly contrasted with clusters of grapes.The eggs go through the limited neck of the follicle into the empty inside the ovary, whence they travel forwards to the oviducts. The oviducts are straight wide tubes, which join to frame so-called uterus over the vagina. The vagina is fixed with the chitin of the body-wall, and leads specifically to the epigyne above described. Opening laterally out of the vagina are two narrow ducts prompting the spermathecae in which the spermatozoa got from the male spider are stored until the eggs are laid. In a few spiders this is the main access to the spermathecae, in others there are free openings to the outside constituting some portion of the epigynum. Spermathecal glands may likewise be present (Savory, 1928).

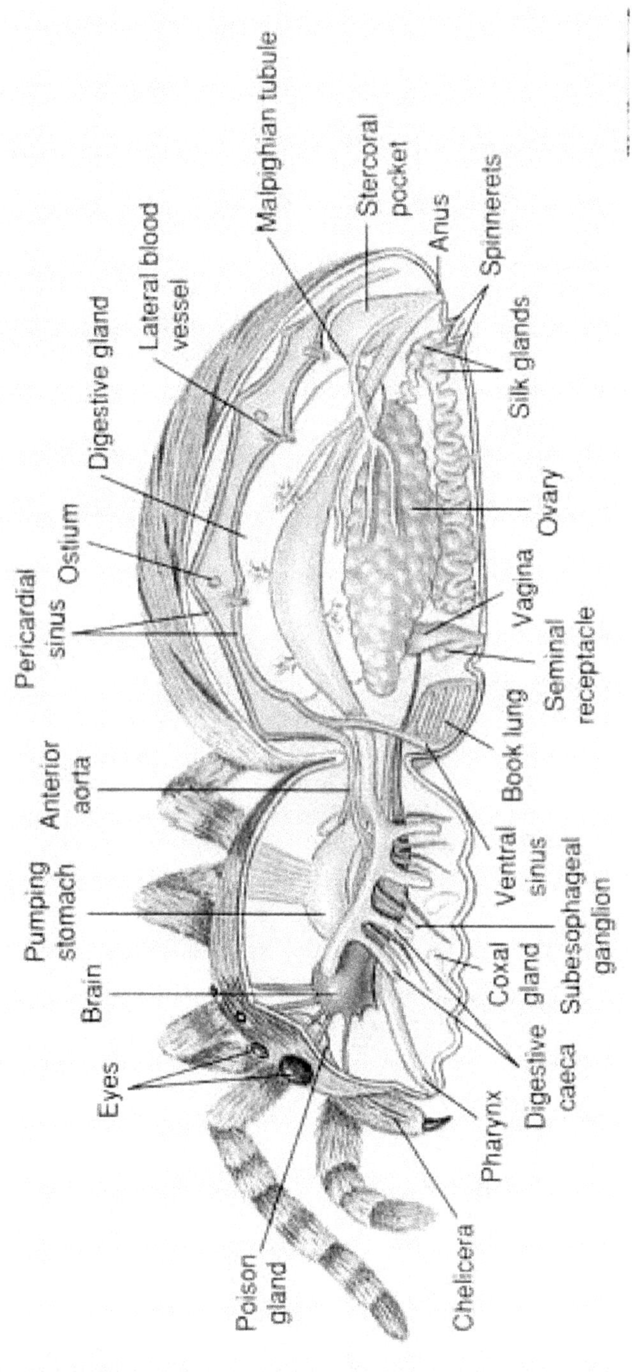

Plate-12: Anatomy of Spider

Silk Production and Webs

The Silk Glands

Silk is utilized for a few unique purposes by various animals, and it is delivered from various parts of their bodies. Caterpillars, for instance, turn silk from an altered salivary organ close to the mouth, and antlions from an adjusted malpighian tube close to the anus, while the silk glands of spiders are, as of now proposed, most likely changed coxal glands of abdominal appendages. Plainly, in this way, silk glands in the distinctive order of Arthropoda are not identified with each other; the silk-creating propensity has emerged autonomously in the few groups. The silk glands of spiders are, as may effortlessly be imagined, of extensive complexity in animals whose lives rely upon their functions. Seven various types of glands are to be discovered having orifices on the spinning organs. These are:

1. The Aciniform glands
2. The Pyriform glands
3. The Ampullaceal glands
4. The Cylindrical or Tubuliform glands
5. The Aggregate glands
6. The Lobed glands
7. The glands of the cribellum

No spiders has each of the seven sorts of glands, yet all have the initial three. The round and hollow glands are controlled by every single female spiders aside from those of the families Dysderidae and Salticidae. The total glands are discovered just in the three most exceptionally specific families, the Theridiidae, the Linyphiidae, and the Epeiridae; and the Theridiidae alone have lobed organs. Ultimately, the cribellum glands are, obviously, discovered just in association with that organ (Underwood notes).

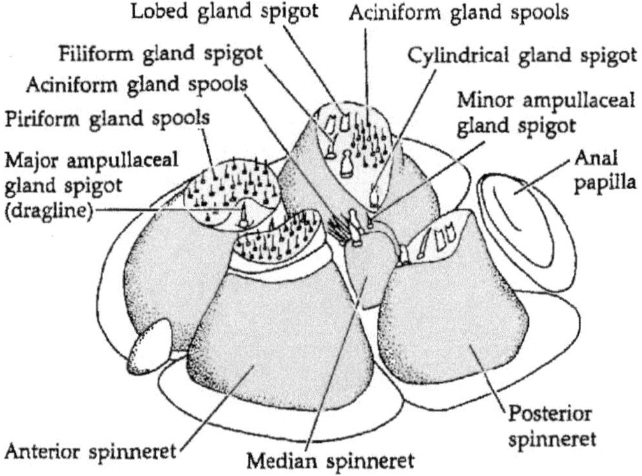

Plate-13: Spinneret organ of spider for silk production

Silk is a fibrous protein made out of the amino acids glycine, alanine and serine. It is delivered in a fluid, water solvent frame that dries into an insoluble shape as it leaves the body. Silk is delivered by uncommon organs and is discharged through specific structures called spinnerets.

The main abdominal appendages persevering in the grown-up spider are those of the fourth and fifth segments, where they work as the spinning organs—to be specific, the cribellum, where this organ is available, and the six spinnerets. The cribellum represents the endopodites of the fourth segment, whose exopodites are the foremost or prevalent spinnerets. The little center spinnerets are the endopodites of the fifth, and the exopodites of this segment are the posterior or inferior spinnerets. Each spinneret is composed of a mix of little and huge tubes called spools and spigots, which open to the outside. The distinctive spools and spigots enable the spider to spin silk into threads of various thickness. The spinnerets are believed to be gotten from opisthosomal appendages as they have rather complex musculature. There are around six diverse silk glands creating six various types of silk (Underwood notes).

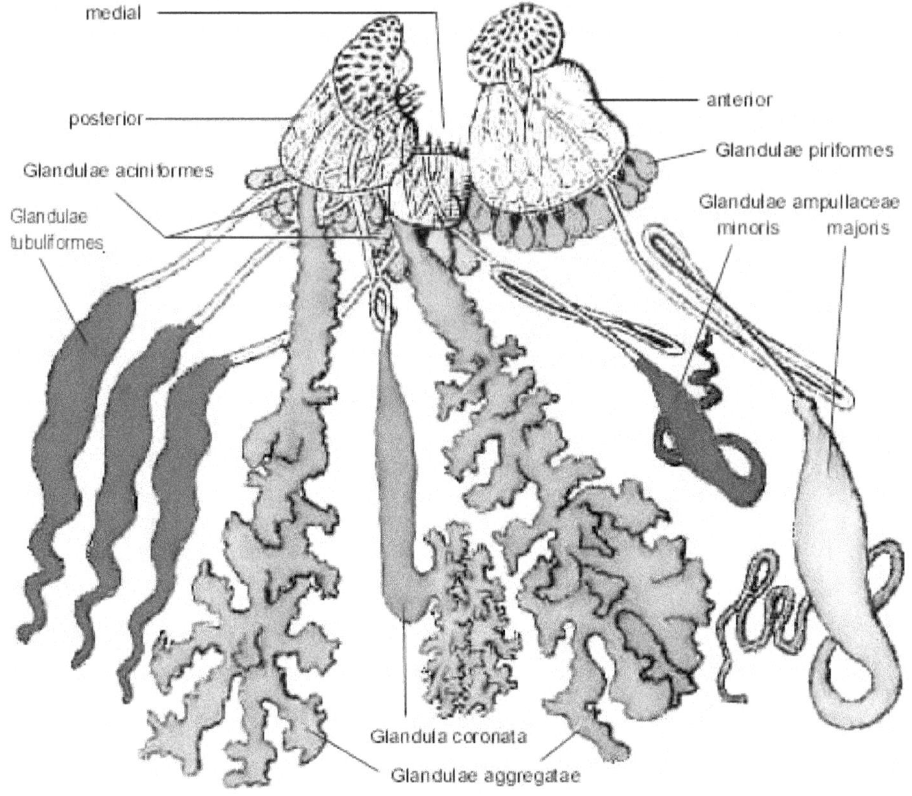

Plate-14: Spider's silk gland

These distinctive sorts can be entwined to create an assortment of silk strings. A few spiders likewise have a cribellum, a spinning organ that endures to 40,000 spigots that create silk in to a great degree fine strings that are brushed by the spiders utilizing specific structures on the fourth walking legs called calamistra. Silk is utilized as security lines and climbing lines, for the development of nests, cocoon, and traps, to wrap prey for storage, as egg sacs and sperm platform, and to line tunnels. Youthful spiders additionally create parachute-like structures to ride the wind as they scatter from their natal zone. At long last silk is utilized as a part of prey catch in development of webs (Underwood notes).

Silk production continues: An example of web construction

A standout amongst the most surely understood kinds of web is the sort developed by orb weaver spiders. The spiders at first spin a string that is conveyed by the wind until the point that it appends to something. The spiders anchors its side of string, moves to the focal point of the line and drops, shaping a Y. Radial strings are then set connected to outline strings and a brief winding is spun. The second phase of development includes the situation of the sticky winding, which is constantly covered with stick that takes after numerous little beads along the strings. Some orb weavers additionally create a thick mesh of silk called the stabilimentum. It is suspected that this alarms feathered creatures to the area of the web with the goal that they won't fly into it causing harm. The genuine capacity is still wrangled about (Underwood notes).

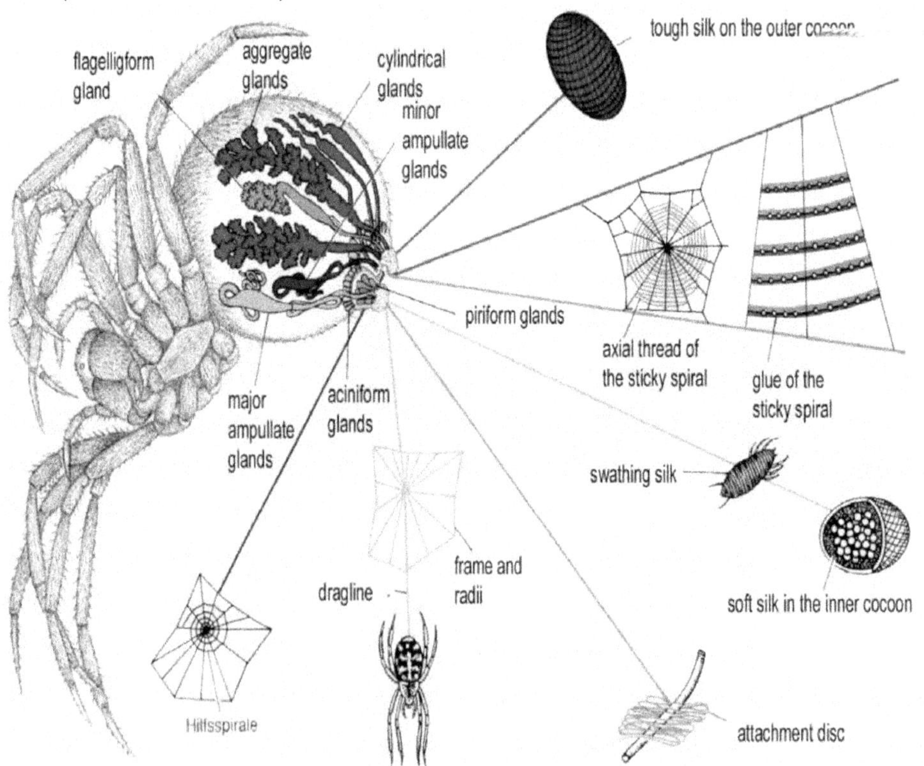

Plate-15: Types of silk produced by spider's through silk glands

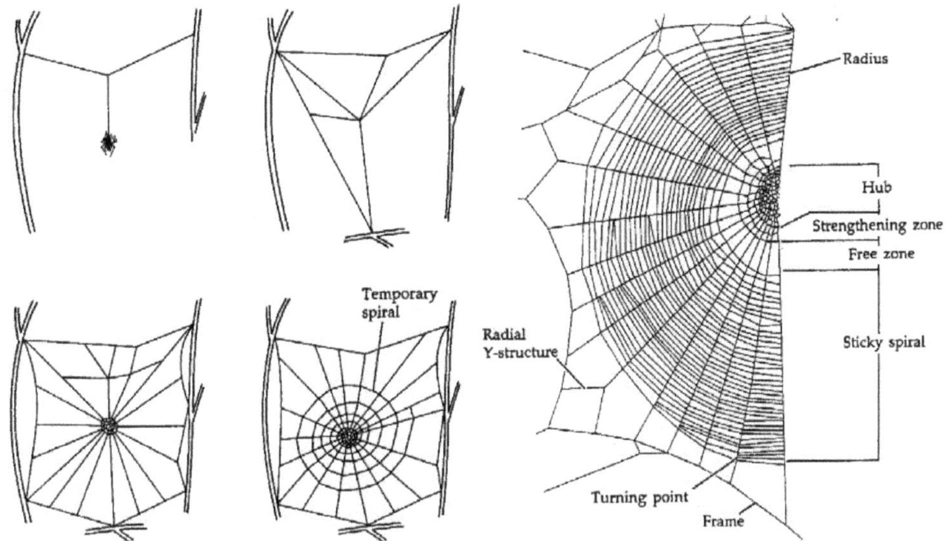

Plate-16: Web formation by spider

Sensory Structures

The majority of the "hairs" on a spiders are mechanoreceptors. A few spiders likewise have thin material hairs on the legs called trichobothria (plate-17), which are greatly sensitive to even airborne vibrations (beating bug wings, air streams, sound frequencies, and so forth.). Chemoreceptors are discovered related with little hairs encompassing the mouth, on the pedipalps, and are most bottomless on the tips of the legs where contact with the substrate is made. Vision is generally insignificant for spiders that rely on catching prey inside webs. These spiders utilize pattern of vibration on the web to recognize among prey, predator, or mate. Among the spiders that catch prey without webs, vision is exceedingly imperative. spiders have single-lensed rhabdomeric eyes. The individual sensory units are basic ocelli in a bunch. The principle eye has the light-sensitive portions of the sensory cells coordinated toward the focal point. The secondsry eyes are inverted with the light-receptor components coordinated far from the focal point. The secondary eyes are thought to serve under low light conditions (Underwood notes).

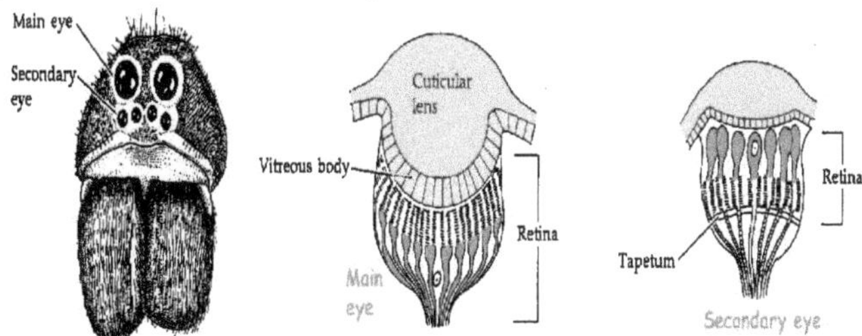

Plate 17: Spider's sensory structures

(Source-www.csulb.edu/~dlunderw/entomology/4-Araneae.pdf)

Prey Capture by Spiders

About all spiders deliver proteinaceous neurotoxins in the toxic substance secreted by glands related with the fangs of the chelicerae. There are essentially two kinds of spiders, those that catch their prey utilizing a silken web or trap and those that chase down their prey and catch them without the utilization of silk (Underwood notes). Beside the ordinary online prey catch, various spiders utilize silk to get prey in rather novel ways.

1. A sheet web is delivered by turning a dome shaped sheet bolstered by a system of strings. Insects get trapped by the supporting strings. The spiders shakes the whole web, thumping the insects into the sheet underneath where it is pulled through the web by the spider.
2. Bolas spider utilize sticky strings to throw onto prey to catch them much like little lassos.
3. Funnel web spiders fabricate a web taking after a funnel; the spiders can be found at the base of the web in the hole.
4. A crude group of spiders lives in a tunnel and trails "fishing lines" or outing strings to alarm the spider when a prey item is adjacent.
5. Purse web spiders spin a silken tube that lies upon the ground. The spider covers up inside and when it identifies the vibrations of go after the surface it slices through the webbing with its chelicerae and catches the prey.
6. Net-casting spiders utilize a little web that they throw onto their prey much like throwing a net for fish.

Spiders that are comparable to cheetahs incorporate the lycosids and the salticids. They rundown their prey or lie in snare. Crab insects regularly are disguised and catch insects that visit blossoms (Underwood notes).

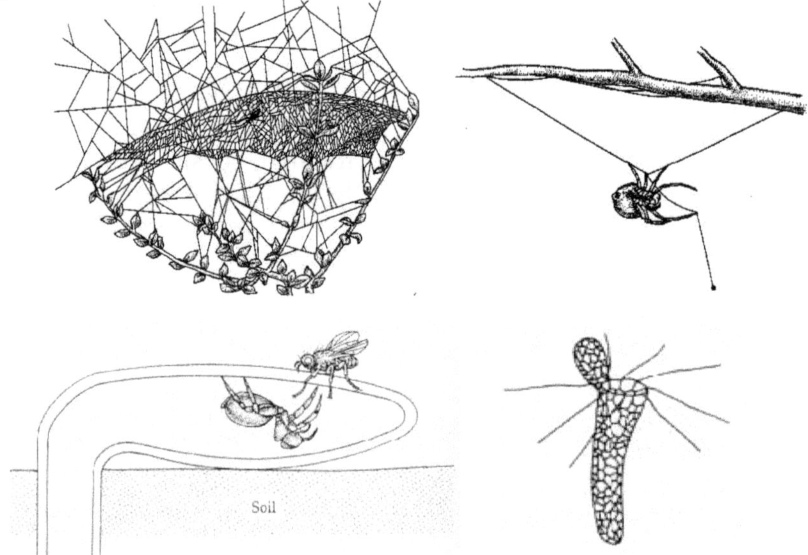

Plate-18: Prey capturing by spiders (source-www.csulb.edu/~dlunderw/entomology/4-Araneae.pdf)

Considering spiders are predators, courtship and mating can be a perilous recommendation. On the off chance that a potential mate is misidentified as either being prey or predator, passing might be all the cheerful suitor finds. Spiders have created complex components to guarantee mating happens without damage. Every male insect utilize their pedipalps to exchange a spermatophore to the female. The spermatophore is bundled in silk. Most males demonstrate particular alterations to pedipalps related with mating. Courtship and species-particular conduct and mating positions additionally help safeguard mating is with conspecifics (Underwood notes).

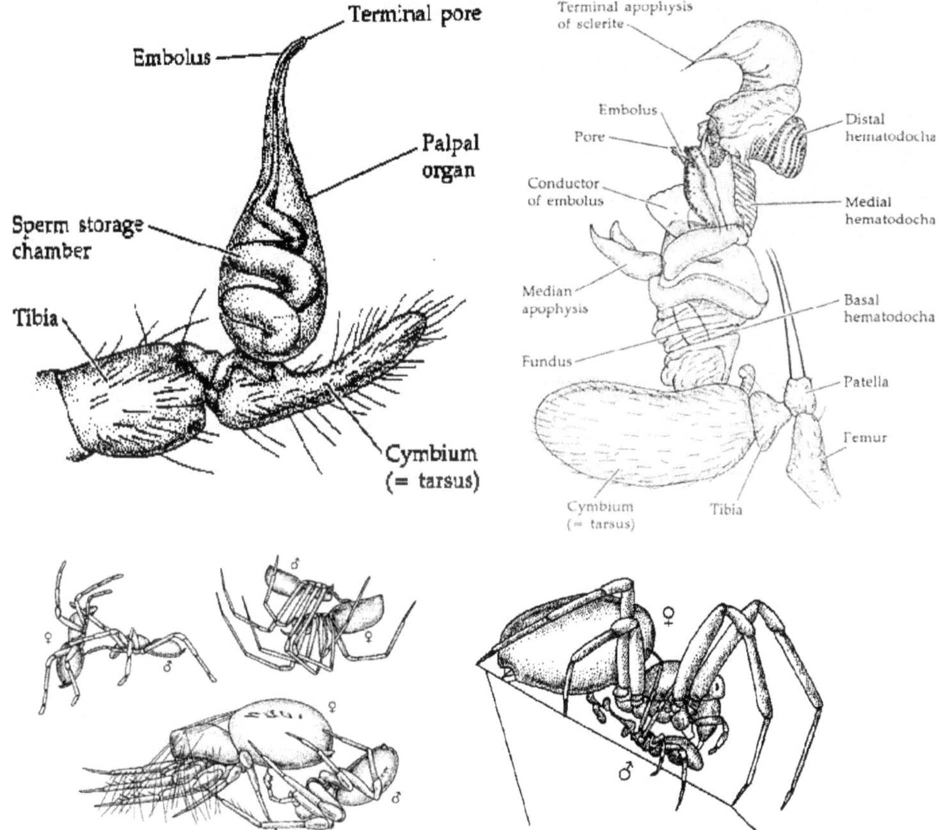

Fig. 11: Courtship, mating, and reproduction (source- www.csulb.edu/~dlunderw/ entomology/4-Araneae.pdf)

References

Underwood, D.L.A. **Biology of spiders.** Biology 316- General Entomology. www. csulb.edu/~dlunderw/entomology/4-Araneae.pdf.

Savory, H.T. (1928)- The biology of spiders, London SIDGWICK & JACKSON, LTD.

General information on spiders- http://www.vapaguide.info/page/22

Spider morphology- https://academics.skidmore.edu/wikis/NorthWoods/index. php/ Spider_morphology

Chapter 4

Suppresion of Rice Brown Plant Hopper by Spiders

Mandeep Kaur, Amandeep Singh, Randeep Singh and Kiranjot Kaur

All the organisms in a rice field are involved in interactions within and between species, including prey-predator and host-parasite relationships (Hijii, 1984). In particular, the interaction of prey and predator show a constant numerical interaction. Information about these relationships is fundamental to biological control. Yamano (1977) suggested that spiders are the most important biological control agents regulating insect populations in rice fields, including insect pests. Predators tend to cluster in stable sites where many prey species are maintained at a high density. Seasonal fluctuations in the numbers of spiders in rice fields showed a constant numerical interaction with their prey i.e. insect pests.

The brown planthopper (*Nilaparvata lugens* Stål) is a pest which feeds only on rice plants (Otake and Hokyo, 1976). Insecticides have been widely used to control brown planthoppers in many Asian countries (Nagata, 1985). Integrated pest management program is one of the strategy that use natural enemies to control planthoppers. Many studies have reported numerous parasitic and predatory natural enemies of planthoppers. There are 14 species of spiders belonging to 13 genera in Korea (Okuma, 1958). For effective control of brown planthoppers, predators should inhabit the same part of the rice plant, to reduce the time spent searching for prey. Brown planthoppers prefer to live on rice plants about 10 cm above the surface in the irrigated field (Lin, 1970). This part of the plant is usually shaded and has a high relative humidity and a relatively low temperature, both of which are favorable for reproduction of the pest. Leasar and Umzicker (1978) documented that humidity and temperature are the most important factors affecting the distribution of spiders. The dominant species of spiders in live on the lower 20 cm of rice plants.

Among the 160 reported species of invertebrate predators of planthoppers, spiders represent the dominant group (Chiu, 1979). Similarly, Lee *et al.* (1997) documented that more than 90% of natural enemies of brown planthoppers are spiders in paddy fields of Korea. Both hunting spiders (Lycosidae) and web-builders (Linyphiidae and Tetragnathidae) were the most important predators of planthoppers (Kenmore, 1980). Among twelve species of natural enemies found in these field experiments, two species of spiders were the most common: *Pardosa pseudoannulata* Boesenberg et Strand (Lycosidae) and *Araneus inustus* L. Koch (Araneidae). Among the total sampled natural enemies, 60% were *A. inustus* and 30% were *P. pseudoannulata*. In a pure species environment of *A. inustus* or *P. pseudoannulata* can control BPH populations when the predator/prey ratio was lower than 1:18. But at a predator/prey ratio of 1:18 or higher cause mortality lower than 100%. In cage experiments at 1:10 predator/ prey ratio *P. pseudoannulata* killed all BPH nymphs/ day at a ratio of l:50. The BPH adult mortality/day was upto 46% (Dyck and Orlido, 1977).

Table 2. Relative toxicity of insecticides to the spider, *Pirata subpiracticus* and the brown plant hopper, *N. lugens*

Insecticides	Formulation	Number of Tested	Mortality		Relative Toxicity
			*P. subpiraticus**	*N. lugens***	
BPMC	Emulsion	60	86.7	100	0.867
BPMC	Dust	60	100	100	1
Carbofuren	Granules	60	100	100	1
Tebufenozide with BPMC	Wettable powder	60	70	100	0.7
Imidacloprid	Spray	60	16.7	100	0.167
Bufrofezine with Isoprocarb	Wettable powder	60	90	100	0.9
Deltamethrin	Emulsion	60	96.7	100	0.967

± 24 °C, *RH-60 %, In a pot (7cm in diameter) in which rice was planted and covered with a glass lid*

A/B< 1 (Effective to N. lugens), A/B >1 (Effective to P. subpiratus) and A/B= 1 (Effective to both)

**Subadult (7th to 9th Instar), ** Adult*

Source- (Joon-Ho Lee and Seung-Tae Kim 2001)

In mixed prey environments, spiders choose prey that is easiest to catch and eat and target prey mortality may be smaller. In ambient field conditions (temporary cages), the mortality of BPH was lower when mixture of prey species was available as compared to pure species environment. With a predator/prey ratio of 1:3 to 1:11, these two spider species caused 78 to 91% of BPH mortality when prey population was in normal condition (i.e. non-outbreaks). Similarly, Stapley (1976) reported that population of *N. lugens* in rice field can be controlled by *P. pseudoannulata* at predator/prey ratio of 1:4-1:8. The Lycosa spider ratio of 1:8 can control BPH populations at non-outbreak level. According to Preap (2001), *P. pseudoannulata* and *A. inustus* (30 and 60% population respectively of natural enemies) are common species of species of spiders in Cambodian rice fields and are key factor for control of BPH attacks. As the population of spider increases along with BPH populations, a predator/prey ratio at not higher than 1: 11 is sufficient to save a crop from pest attack. There is need to take care of spiders and others natural enemy in the fields. Field monitoring is recommended to follow the crop situation and determine the predator/prey ratio for decision making in pest control. If a spider/BPH ratio is higher than 1:11 and continues to increase up to or higher than 1:20, then the crop is in danger of damage and a chemical control would be recommended especially for a susceptible variety (Preap *et al.*, 2001).

The relative toxicity to spiders of various insecticides used against brown planthopper was tested by Lee and Kim, 2001 given in Table-2. It was reported that granule formulation of insecticides were most lethal to spiders than any other formulations.

Lee and Kim studied use of spiders as natural enemies to control rice pests in Korea. Sub adults of *P. subpiraticus* consumed the maximum number of prey, although the total number of prey consumed by *C. kurilensis* and *G. dentatum* were as high. Spiders belonging to the species *P. subpiraticus* showed the greatest variation in daily consumption, while the number of planthoppers consumed/ day by *C. kurilensis* and *G. dentatum* varied by only 0-4 planthoppers. It appears that the difference in the numbers of planthoppers consumed by hunters and by web builders was due to the differences in the efficiency of their foraging strategies. Generally, female spiders consumed more prey than males, except *G. dentatum*. It is thought that females usually need more energy for oviposition and brood care, while males need only the energy for survival. Sub adults had a higher consumption rate than adults, which implies that the energy needs for growth are higher than those needed for reproduction.

References

Hijii, N, (1984). Arboreal Arthropod Fauna in a Forest. II. Presumed Community Structures based on Biomass and Number of Arthropods in a Chamaecyparis Obtusa Plantation. *Japanese Journal of Ecology* **34:** 187-93.

Yamano, T. (1977). Seasonal fluctuation of population density of spiders in paddy field in Kyoto City *Acta Arachnology* **27:** 253-60.

Otake, A. and Hokyo, N. (1976). Rice Plant and Leafhopper Incidence in Malaysia and Indonesia. Report of a Research Tour January to March 1976. Shiryo No. 33, Tropical Agriculture Research Center, Tokyo. 64 pp.

Nagata, T. (1985). Chemical control of the brown planthopper in Japan. *Japanese Agricultural Research Quarterly* **18:** 176-81.

Okuma, C. (1958). Some observations on the habits of Oedothorax insecticeps Bösengerg et Strand (Araneae) *Acta arachnology* **15:** 21-23.

Preap, M. P., Zalucki, G. C., Jahn, L. and Harry, J. N. (2001). Effectiveness of Brown Planthopper Predators: Population Suppression by Two Species of Spider, Pardosa pseudoannulata (Araneae, Lycosidae) and Araneus inustus (Araneae, Araneidae) Visarto *J. Asia-Pacific Entomol.* **4 (2)** : 187 -193

Lin, K. S. (1970). Studies on the microclimatic factors in relation to the occurrence of the rice planthopper. *Plant Protection Bulletin* **12(4):** 184-89.

Leasar, C. D. and Umzicker, J. D. (1978). Life history, habits and prey preferences of Tetragnatha laboriosa (Araneae: Tetragnathidae) *Environmental Entomology* **7:** 879-84.

Chiu, S. C. (1979). Biological control of the brown planthopper. In: Brown Plant-hopper, Threat to Rice Production in Asia. International Rice Research Institute, Los Baños, Laguna, Philippines, pp. 335-55.

Lee, J. H., Kim, K. H. and Lim, U. T. (1997) Arthropod community in small rice fields associated with different planting methods in Suwon and Icheon. *Korean Journal of Applied Entomology* **36(1):** 55-66.

Kenmore, P. E. (1980). Ecology and outbreaks of a tropical insect pest of the green revolution, the rice brown planthopper, Nilaparvata lugens Stal, Ph. D. Dissertation. University or California at Berkeley, CA, U.S.A.

Dyck, V. A. and Orlido, G. C. (1977). Control of the brown planthopper (Nilaparvata lugens Sial) by natural enemies and timely application of narrow-spectrum insecticides. Pp. 58-72_ In the Rice brown planthopper, Food and fertilizer Technology Center 1'01' the Asian and Pacific Region. Taipei, Taiwan.

Stapley, I. H. (1976). The brown planthopper and Cyrtorhynus spp. Predators in the Solomon Island Islands. *Rice Entornol, Newsl.* **4**:17

Chapter 5

Ecological Importance of Spiders

Gurbax Singh, Amandeep Singh, Mandeep Kaur and Kiranjot Kaur

Spiders as Predators in Rice Field

Agricultural entomologists recorded the importance of spiders as a major factor in regulating pest and they have been considered as important predators of insect pests in agroecosystem. They serve as a buffer to limit the initial exponential growth of prey population (Synder and Wise, 1999). However, researchers have exposed that spiders in rice field can play an important role as predators in reducing planthoppers and leafhoppers (Chiu, 1979). It is reported that about 30 white leaf hoppers can be consumed by a single spider in a day. Under favorable conditions, these can reach maximum densities upto 1000 indiviuials/ m^2 approximately (Pearse, 1946). Spiders belong to class arachnida in which head and neck are joined. Use of spiders against insect pest having one advantage over insect predators is that it only feed on insect not on rice, whereas other insect predators feed both on rice and insects. Currently 39,882 valid described species of spiders in 3676 genera and 108 families have been described globally (Platnick, 2011). The spider fauna of India is represented by 1520 spider species belonging to 377 genera and 60 families (Sebastion and Peter, 2009). All spiders are predaceous and play an important role in reduction of pest population because of their abundance and high predatory potential. Most of them are polyphagous predators, able to feed on various insects including major rice pests (Ito *et al.*, 1962). In addition to killing pest by direct attack, they also cause pest mortality by dislodging them from the plants or trapping them in the webs (Nyffeler *et al.*, 1994a). Family of spiders that are often found in agro-ecosystems and play an important role in the control of insect pest species are members of the Araneidae, Linyphiidae, Lycosidae, Oxyopidae, Salticidae, Tetragnatidae, and Thomisidae (Susilo, 2007).

The spiders in lycosidae group usually found in packs of 3 or 4 like wolves and they often called wolf spiders e.g. *Lycosa* spp., *Pardosa* spp., *Wadicosa* spp. In thomisidae group actually the spiders look like crabs and can move in all directions without turning around. They are often called crab spiders e.g. *Thomisus* spp. Tetragnatidae spiders are well developed and with long body, they are giant spiders and sometimes known as 'long-jawed' spiders e.g. *Tetragnata* spp. The spiders of Araneidae or Argiopidae are the best weavers and weave spider webs e.g. *Argiope* spp., *Gasteracantha* spp. The legs of Pholcidae spiders are very long and thin and often called as grand-dad long-legs *e.g. Pholicus* spp. Salticidae spiders are also known as jumping spiders and these spiders more prefer to leap and jumping instead of running and walking *e.g. Salticus* spp.

Plate-19: *Lycosa* spp. *Spiders*

Plate-20: *Pardosa* spp. spiders

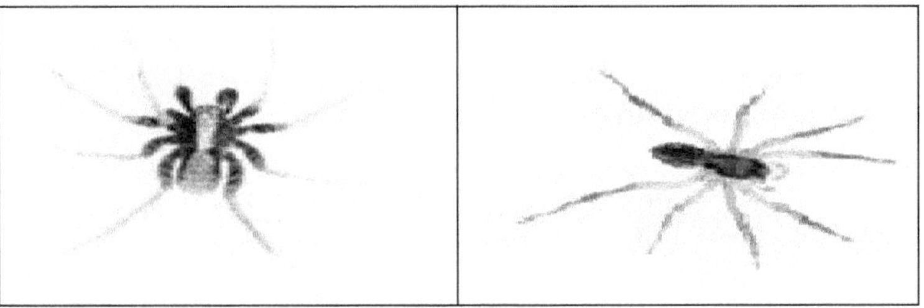

Plate-21: *Wadicosa* spp. Spiders

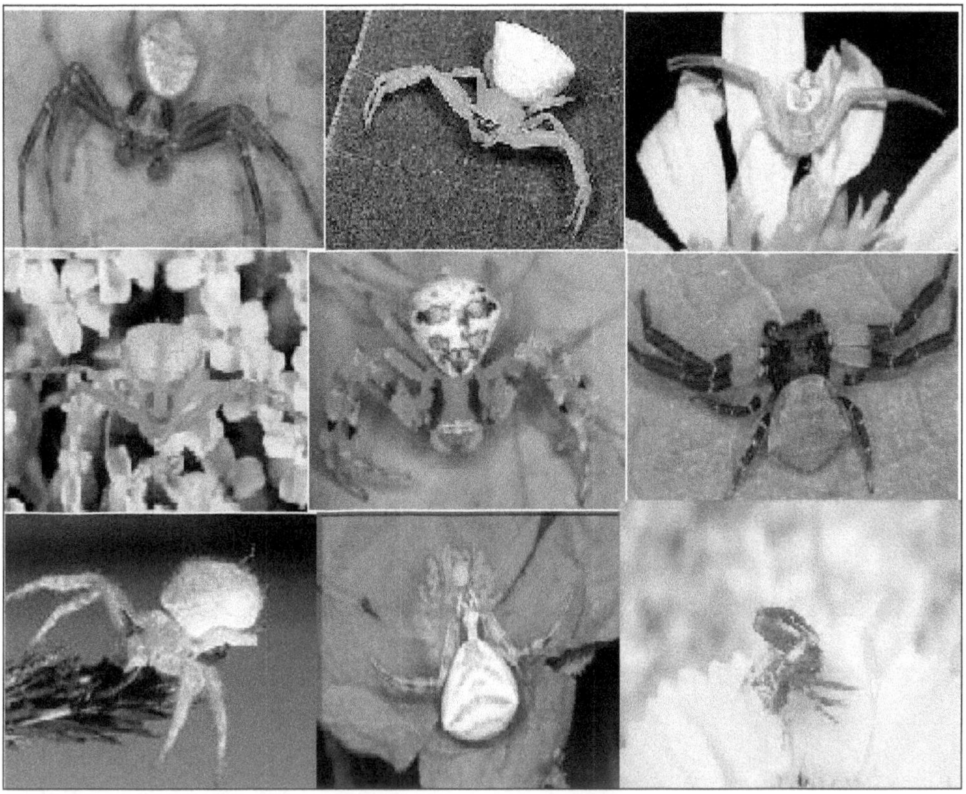

Plate-22: *Thomisus* spp. spiders

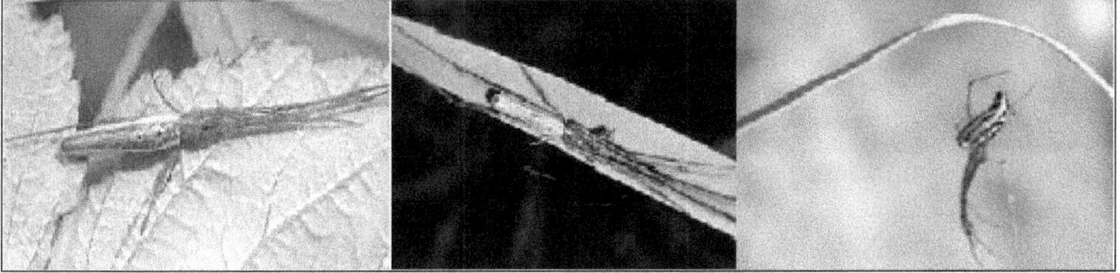

Plate-23: *Tetragnata* spp. Spiders

Plate-24: *Argiope* spp. spiders

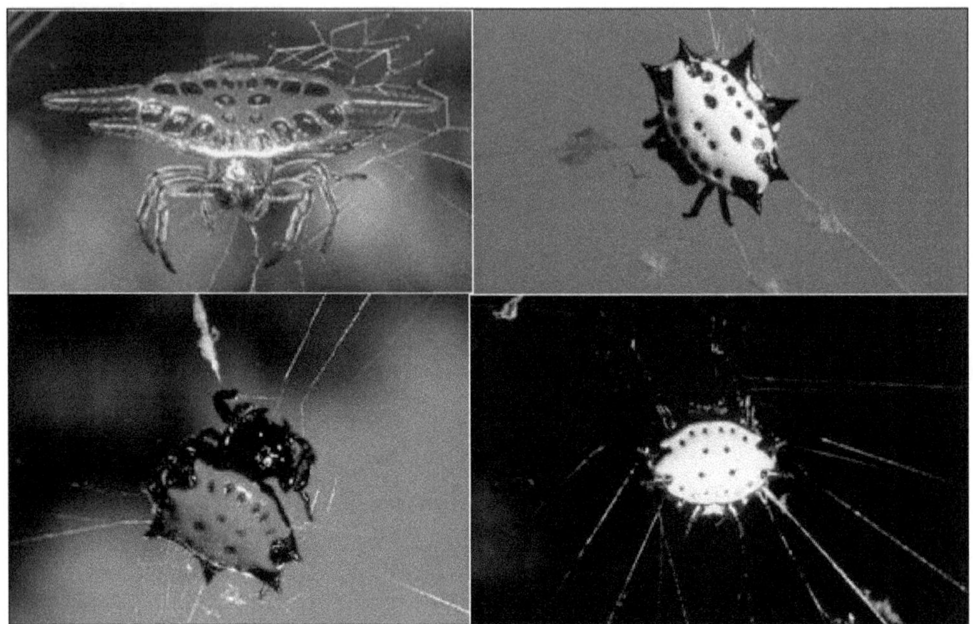

Plate-25: *Gasteracantha* spp. spiders

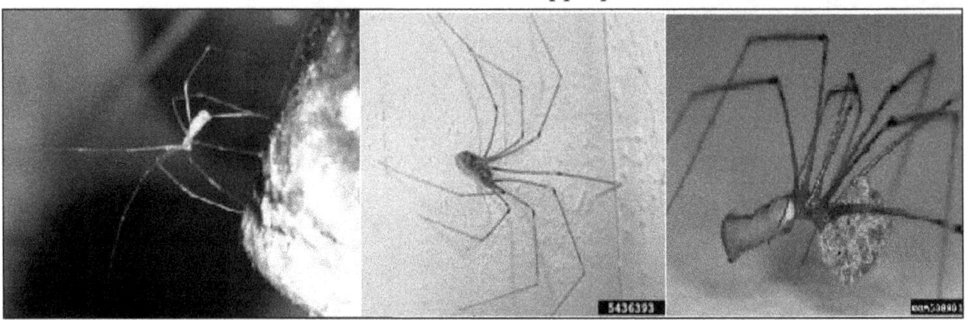

Plate-26. *Pholicus* spp. spiders

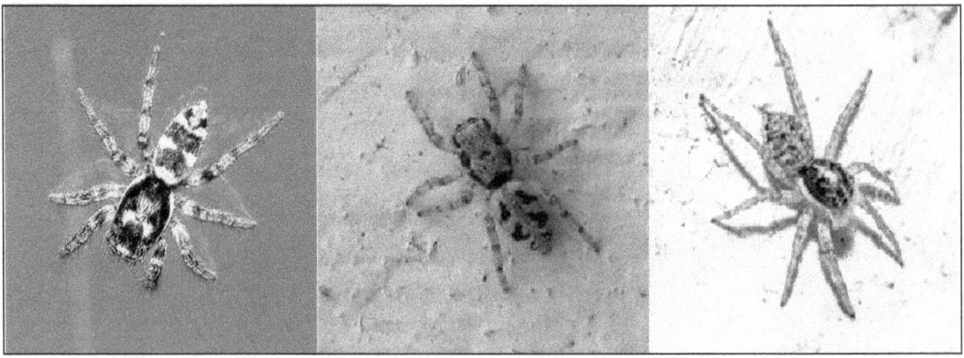

Plate-27. *Salticus* spp. spiders

Types of Spiders

Usually three groups of spiders are found in rice field. These are orbweaving spiders, hunting spiders and space-web spiders. Orb-weaver spiders are included in the families i.e. Araneidae, Tetragnathidae and Theridosomatidae. Tetragnatha, Araneus and Argiope are the most common genera of orb-weaver spiders. The orb-web weavers Araneidae and Tetragnathidae feed upon pest belonged to order Homoptera such as leafhoppers, Diptera and Orthoptera especially grasshoppers (Sigsgaard, 2000). Orb-weaver usually becomes abundant when insect damage has already occurred (Barrion and Litsinger 1984). Hunting spiders belong to families i.e. Lycosidae, Oxyopidae, Thomisidae and Salticidae and frequently capture insects of order Orthoptera, Homoptera, Hemiptera, Lepidoptera, Thysanoptera, Diptera and some Coleoptera. In the group of hunter spiders Lycosids dominate (Nyffeler *et al.*, 1994a). The space-web weavers belong to families i.e. Linyphiidae, Dictynidae and Thridiidae and capture dipteran, Hemipteran and Homopteran (especially aphids and leafhoppers) pests as well as beetles in the family curculionidae (Barrion and Litsinger, 1995).

Sau *et al.*, (1996) reported that after first 35 days of transplantation of rice lycosid *Pardosa pseudoannulata* and linyphiid *Atypena formosana* are the dominant predators in irrigated rice field. While three families included in space-web spiders group are Theridiidae, Linyphiidae and Agelenida (Barrion and Litsinger 1995). In the cropping season orb-weaving spiders are the most abundant spiders, with *Tetragnatha* spp. being the single most common genus in South East Asian countries, except *P. pseudoannulata* which is most dominated species in Philippines. Across the season, the relative abundance of *P. pseudoannulata* was reported of 25 to 54% of all spiders at five rice sites in the Philippines (Heong *et al.,* 1992). Maximum abundance of *A. formosana* was found at the lower elevations comprises 23, 35 and 40% and lowest at the two sites of higher elevations of 800 and 1500 m (7 and 9%). In these sites, three species of tetragnathids together comprised 10 to 39 % of the spiders. These are *Tetragnatha virescens* Okuma, *T. Maxillosa* Thorell and *T. javana* Thorell. The occurrence of these spiders remains throughout the year.

Barrion and Litsinger (1984) reported that *P. pseudoannulata* having wide prey diversity which including leafhoppers *P. pseudoannulata* is most common among the tillers at the base of the plants. It preys on a wide array of insect pests, including leafhoppers and planthoppers, whorl maggot flies, leaffolders, caseworm and stem borers (Barrion and Litsinger 1984). Field densities of both spiders co-vary with hopper densities (Reddy & Heong 1991). References to the importance of the smaller and less conspicuous *A. formosana* have been few until recently (Shepard *et al.,* 1987). *A. formosana* adults and immatures prefer to live among the rice stem or at the base of rice hills. They have been observed to hunt for nymphs of planthoppers and leafhoppers, Collembola, and small dipterans, such as whorl maggot flies (Barrion and Litsinger 1984). Later in the cropping season predatory bugs become the most numerous predators. The most abundant of these are *Microvelia douglasi atrolineata* Bergoth (Veliidae), *Mesovelia vittigera* (Horvath) (Mesoveliidae), and *Cyrtorhinus lividipennis* Reuter (Miridae) (Heong *et al.,* 1991).

Wolf spider, *Lycosa pseudoannulata* (http://rkb.irri.org.)

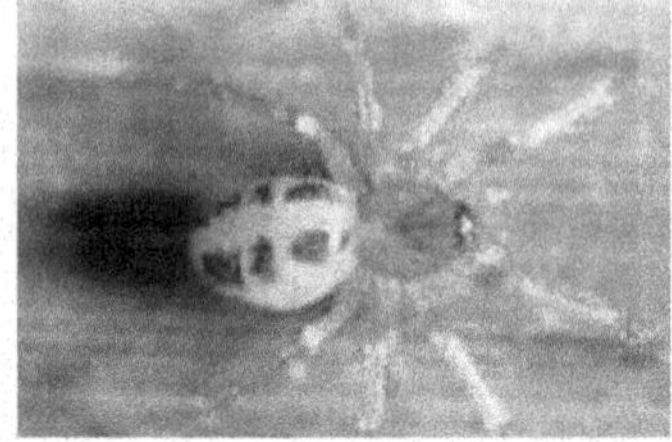

Dwarf spider, *Anyphena formosana* (htpp://rkb.irri.org.)

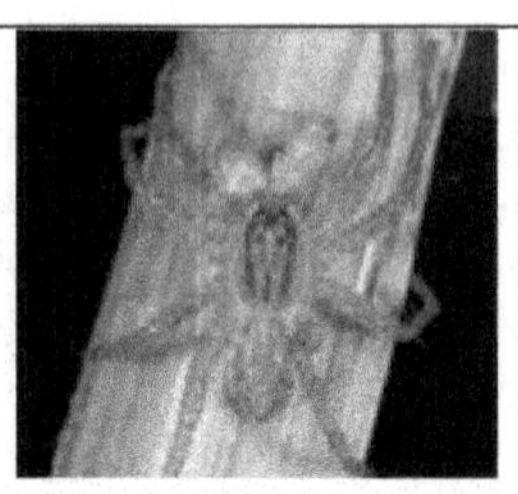

Lycosid spider, *Pardosa pseudoannulata* feeding on BPH (http://rkb.irri.org)

Source: *http://www.assignmentpoint.com*

Plate-28: Spiders are abundant in rice fields through the world

Although web builders were more numerous than hunters in some areas (Choi and Namkung 1976), seasonal fluctuations in rice field spiders in Korea are thought to depend mainly on the density and species composition of hunting species of spider (Table-3, Park *et al.*, 1972). Kim (1998) pointed out that if spiders are being used as biological control agents, it is very important to understand their life styles. Both web builders and hunters follow a foraging strategy. Song and Lee (1994) suggested that web builders such as *P. clercki* were better able to suppress insect pests than hunters such as *P. subpiraticus*. However, hunters are usually active predators which follow a "pursue and kill" foraging strategy, while web builders follow a passive "sit and wait" strategy (Enders, 1974). Furthermore, the webbing sites of web builders are easily affected by environmental factors. In addition, when the web spaces overlap, there is competition with and between species of web builders. Therefore, hunters probably are more effective predators than web builders.

Spider as natural enemies of insect pests

As natural control agents the ecological importance of spiders is to be expected. According to Whitecomb (1974) on ecological point of view spiders play a very important role in diverse respects e.g. they can be food for other predators. Thornhill (1975) reported that dead insects in spider webs can be food sources. Under grassland conditions (where vegetation having without the human interferences) often having high population densities of two weaving spiders (large funnel web spiders and orb-weavy spiders). Nyffeler in 1982 reported that on an average *Argiope bruennichi* (Orb weavers) may capture upto 7 grasshoppers/ web/ day and 6 webs per m^2 in an abandoned grass land near bonn (Federal republic of Germany) reported by Lothmeyer & pretscher in 1979. Nyffeler in 1982 reported that the web area covered by *A. bruebbichi* was 3000 m^2/ha ground area. It is evident that such a large area of web by spider exerts an enormous predatory pressure on insect population. If every spider captures about 10 insects/web/day then in 50 days orb-weavers would destroy 3×10^7 insect/ha (Nyffeler). Kajak *et al* in 1968 demonstrated that under abandoned grassland area orb weaving spider can exert a strong predatory pressure on the insect population. In undisturbed grasslands conditions the web-building spiders eliminated 25-40% of adult diptera.

Source: *en.wikipedia.org*

Plate-29: Orb weaver spider

In undisturbed grasslands, high population densities of funnel web spiders *Genus-Agelena* found. On an average, upto 12 funnel web spiders /m^2 found by Nyffeler and Benz during 1978 in near Zurich (Switzerland) in some abandoned grassland habits. They reported that food of such spiders as composed of upto 23% of bees. By this example, it is demonstrated that spiders can have a noticeable effect on insect population. Van Wook (1971) and Schafer (1974) reported that in undisturbed grasslands not only web-building spiders, but also hunting spider can be impotant natural control agents of invertebrates.

In cultivated meadows and annual crops agro-ecosystems, the periodical harvest of the vegetation is an extreme stress for the foliage-dwelling spiders that is unstable for the colonization by and survival of foliage dwelling spiders (Tischler, 1958). In cultivated meadows and cereal fields the population densities and biomass of spiders are low, the number of prey killed by the foliage-dwelling spiders communities of these cultivated habitats are low as compare to undisturbed grassland areas. It is estimated that in the vegetation of such cultivated meadows and annual crops less or equal to 2 kg insects (fresh weight) /ha/ year are killed by spiders (Kajak, 1971). Near Zurich (Switzerland) in cereals fields together with the carnivorous beetle (Carabidae and staphylindae), they belong to the most abundant ground dwelling predators of insects. Within the ground-dwelling predator complex, German studies had also established that spiders form an important component. Basedow (1973) reported that the ground dwelling predator potentially exerts a strong predatory pressure on insect population. Many studies from different locations of world (Japan, India, China, Taiwan, Korea, Thailand and Phillipines) indicated that ground-dwelling spiders have a damping influence on the population densities in rice fields (Okuma, 1968). Miller and Obrtel in 1975 reported that spiders play an important role in European swap ecosystem. Due to swap ecosystem of rice fields spiders play a major role as predator for insect pest.

References

Barrion, A. and Litsinger, J. (1995). Riceland spiders of South and Southeast Asia. CAB International/International Rice Research Institute. University Press, Cambridge.

Barrion, A.T. and Litsinger, J.A. (1984). The spider fauna of Philippine rice agro ecosystems. II. Wetland *Philippine Entomologist* **6:** 11-37.

Choi, B.R., Heong, K.L., Lee, J.O., Yoo, J.K. and Park, C.G. (1996). Effects of Sublethal Doses of Insecticides on the Brown Planthopper, Nilaparvata lugens Stål (Homoptera: Delphacidae) and Mirid predator, Cyrtorhinus lividipennis Reuter (Hemiptera: Miridae) *Korean Journal of Applied Entomology* **35(1):** 52-57.

Choi, J.S., Goh. H.G., Uhm. K.B., Choi, K.M. and Hwang, C.Y. (1992). Life cycle of the Mirid predator, Cyrtorhinus lividipennis Reuter (Hemiptera: Miridae) *Korean Journal of Applied Entomology* **31(4):** 329-37.

Choi, K.M. and Lee, M.H. (1990). Use of Natural Enemies to Control Agricultural Pests in Korea. In: Use of Natural Enemies to Control Agricultural Pests. Food and Fertilizer Technology Center for the Asian and Pacific Region, Taipei, Taiwan, ROC pp. 40-50.

Choi, S. S. and Namkung, J. (1976). Survey on the Spiders of the Rice Paddy Field (I*) Korean Journal of Plant Protection* **15 (2):** 89-93.

Chua, T. H. and Mikil, E. (1989). Effects of prey number and stage on the biology of Cyrtorhinus lividipennis (Hemiptera: Miridae): A predator of Nilaparvata lugens (Homoptera: Delphacidae) *Environmental Entomology* **18:** 251-55.

Dobel, H.G. (1987). The role of spider in the regulation of salt marsh plant hopper populations. M. S. Thesis, Department of Entomology, University of Maryland at College Park, MD, U.S.A.

Enders, F. A. (1974). Vertical stratification in orb web spiders (Araneae: Araneidae) and a consideration of other methods of coexistence. *Ecology* **55:** 317-28.

Hamamura, T. (1971). Ecological studies of Pirata subpiraticus (Bös. st Str. 1906) (Araneae: Lycosidae) I *Acta Arachnology* **23:** 29-36.

Heong, K. L., Aquino, G. B. and Barrion, A. T. (1991). Arthropod Community Structures of Rice Ecosystems in the Philippines. *Bulletin of Entomological Research* **81(4):** 407-16.

Heong, K., Aquino, G. and Barrion, A. (1992). Population dynamics of plant- and leafhoppers and their natural enemies in rice ecosystems in the Philippines *Crop Protection* **4:** 371-79.

Hijii, N. (1984). Arboreal Arthropod Fauna in a Forest. II. Presumed Community Structures based on Biomass and Number of Arthropods in a Chamaecyparis Obtusa Plantation. *Japanese Journal of Ecology* **34:** 187-93.

Kajak, A., Andrejewska, L. and Wojcik, Z. (1968). The role of spiders in the decrease of damages caused by Acridoidea on meadows *Experimental Investigation Ekol Poloska* **16:** 755-64.

Kajak, A., Breymeyer, A., Patel, J. (1971). Productivity investigation of two types of meadows in the Vistula Vallev. *Predatory Arthropods Ekol Poloska* **19:** 223-33.

Kim, H. S., Im. D. J., Choi, D. R. and Ko, H. K. (1991). In: Encyclopedia of Identification of Natural Enemy. Rural Development Administration, Suweon, Korea, pp. 25-52.

Kim, S. T. (1998). Studies on the Ecological Characteristics of the Spider Community at Paddy Field and Utilization of the Pirata subpiraticus (Araneae: Lycosidae) for Control of Nilaparvata lugens (Homoptera: Delphacidae). Unpublished Ph.D. Thesis, Konkuk Univ., Seoul, Korea, pp 90.

Kuno, E. and Dyck, V. A. (1984). Dynamics of Philippines and Japanese populations of the brown planthopper: Comparison of basic characteristics. In: Proceedings, ROC-Japan Seminar on the Ecology and the Control of the Brown Planthopper. National Science Council, Rep. of China, pp. 1-9.

Lohmeyer, W. and Pretscher, P. (1979). Ueber das Zustandekommen halbruderaler Wildstauden-Quecken-Fluren auf Brachland in Bonn und ihre Bedeutung als Lebenseraum fur die Wespenspinne *Nature Landsch* **54:** 253-59.

Manti, L. (1989). The role of Cyrtorhynus lividipennis Reuter (Hemiptera, Miridae) as a major predator of the brown planthoppers Nilaparvata lugens Stal (Homoptera, Delphacidac). Ph.D. Dissertation, University of the Philippines at Los Banos, Philippines.

Nyffeler, M. (1982). Field studies on the ecological role of the spiders as insect predators in agroecosystems. Ph.D. dissertation. Swiss Federal Institute of Technology, Zurich.

Nyffeler, M. and Benz, G. (1978). Die Beutespektren der Nerzspinnen Argiope bruennichi (Scop.), Araneus quadratus Cl. und Agelena labyrinthica (Cl.) in Odlandwieser bei Zürich

Nyffeler, M., Sterling, W. L. and Dean, D, A, (1994), Insectivorous activities of spiders in United States field crops *Journal of Applied Entomology* **118:** 113–28.

Ooi P A (1992) Biology of brown planthopper in Malaysia *Plant Protection Tropical* **9:** 111-15.

Ooi, P. A. and Shepard, B. M. (1994) Predators and Parasitoids of Rice Insect Pests. In: Biology and Management of Rice Insects. E.A. Heinrichs, (ed.). Wiley, New Delhi, India, pp. 585-612.

Park, J. S., Lee, S. C., Lee, B. H., Kim, Y. I, Park, K. T. and Ahn, K. J. (1972). Influences of the insecticides on the rice insect pests. *Research Report R.D.A.*: 146- 69.

Platnick, N. I. (2011). The World Spider Catalog, occurring in a habitat and when all species in a sample are Version. *Revue Suisse de Zoology* **85:** 747-57.

Preap, V. (2001). Outbreaks of Brown Planthopper (Nilaparvata lugens Stal): Pattern and Process. Ph. D. thesis. University of Queensland, Australia.

Sahu, S, Singh, R. and Kumar, P. (1996). Host preference and feeding potential of spiders predaceous in insect pests of rice *Journal of Entomological Research* **20 (2);** 145-50.

Shepard, B. M. and Ooi, P. A. (1991). Techniques for evaluating predators and parasitoids in rice. In: Rice Insects: Management Strategies, E.A. Heinrichs and T.A. Miller (Eds.). Springer-Verlag, Germany, pp. 197-214.

Shepard, B. M., Barrion, A. T. and Litsinger, J. A. (1987). Helpful Insects, Spiders, and Pathogens. International Rice Research Institute, Los Baños, Philippines. 127 pp.

Sigsgaar, L. (2000). Early season natural biological control of insect pests in rice by spiders - and some factors in the management of the cropping system that may affect this control *European Arachnology* 57–64

Snyder, W. E. and Wise, D. H. (1999). Predator Interference and the Establishment of Generalist Predator Populations for Biocontrol *Biological Control* **15:** 283–92.

Suenaga, H. (1963). Analytical studies on the ecology of two species of planthoppers, the white-backed planthopper, (Sogata furcifera Horvath) and the brown planthopper (Nilaparvata lugens), with special reference to their outbreaks. *Bull. Kyushu Agric. Expt. Stn* **8(1):** 1-152.

Sussilo, F. (2007). PengendalianHayatideMe mbe rdayakanM usuhAlami Hama Tanaman. Yogyakarta: Grahallmu

Tanaka, K. (1989). Movement of the spiders in arable land *Plant Protection* **43(1):** 34-39.

Thornhill, R. (1975). Scorpionflies as kleptoparasites of web-building spiders Nature **258:** 709-11.

Ven Hook, R. L. (1971). Energy and nutrient dynamics of spider and orthopteran populations in a grassland ecosystem *Ecological Monograph* **41:** 1-26.

Chapter 6

Ecology of Spiders in Rice Fields

Amarinder Singh, Amandeep Singh, Mandeep Kaur
and Gurbax Singh

A. Diversity

Diversity of rice is unique among agroecosystems. According to Rypstra *et al.* (1999), spider assemblage density and diversity are closely related to structural complexity of environment, which may be increasing as plants become larger and more complex. Early in the cropping season, flooding of the field stimulates the activity of dipterous insects, which are an important alternative prey for spiders. As the season progressed, the main prey of wolf spiders in rice field is switched from dipterous insect to hoppers (Settle *et al.*, 1996). Thus, the spiders benefit from an early bang of food that helps to build up their populations before primary insect pests become abundant.

Various spider groups feed primarily on the same orders, but in different proportions. Dietary mixing seems to be advantageous by optimizing a balanced nutrient composition needed for survival and reproduction of spiders (Uetz, 1990). Predation rate was also found to differ between the sexes. Female spiders consumed more insect pests than the males under similar conditions. Kiritani *et al.* (1970) reported that hoppers accounted for over 80% of the diet of hunting spiders from August to September. Tahir *et al.* (2009) recorded 111 pests and 9 hunting spiders (four *L. terrestris,* two *P. birmanica* and three *O. javanus*) from 1 m^2 of rice field. Average number of pests killed or consumed by these three species (viz., *L. terrestris, P. birmanica* and *O. javanus*) per day was 4, 3 and 5, respectively. Rodrigues *et al.*, (2009) reported that in rice agroecosystem and adjacent environments at different development stages of the crop in Southern Brazil, there were total of 2717 spiders distributed among 15 families, 2138 were young individuals and 579 minority adults. Among minority adults, 318 were females and 261 were males. Overall 85 morphospecies were identified among the adults. Among the recorded 15 spider families more than half of all spiders belonged to Oxyopidae, Araneidae, and Tetragnathidae families. In the rice agroecosystem areas, Araneidae was dominant. In the rice culture, ambushers were also more common; however, orbicular web builders had a higher proportion, mainly due to Araneidae and Tetragnathidae.

Similarly, abundance of these spiders were reported by Corseuil *et al.* (1994b) in rice field. The abundance of these families in rice fields was presumably because rice plants are good spots for web building and also have the high number of prey occurring in agroecosystems. Bambaradeniya and Edirisinghe (2001) also reported orbicular web builders feeding on insect pests of rice in Sri Lanka. The most abundant guild in that case was the irregular web builders, mainly due to Theridiidae and Linyphiidae.

Source: *ricehopper.wordpress.com*

Plate-30: Diversity of spiders in rice field

Studies was conducted by Barrion (2001) in irrigated rice fields sampled for two cropping seasons-dry (January to April) and wet (September to November) and collected 3,098 individuals. It comprises 42 species belonging to 35 genera and 13 families. On the other hand, the collection of spiders throughout the year from the non-rice habitats varies by location. In three sites in San Juan, Batangas, the total collection is 8,870 spiders with 3,829 individuals (39 species, 33 genera and 14 families) in the broadleaf *(Malachra* spp.) and 5041 spiders in the tall grasses of *Saccharum spontaneum* L. (2880 individuals, 50 species, 38 genera and 15 families) and *Paspalum conjugatum* Berg (2161 individuals, 37 species, 28 genera and 13 families). The grand total collection was much higher in Candelaria, Quezon yielding 26,177 spiders in five non-rice habitats, namely, irrigation canal (6,966 individuals, 47 species, 37 genera and 16 families), edge of the bund (6,488 individuals, 62 species, 47 genera and 19 families), coconut (5 ,833 individuals, 48 species, 39 genera and 18 families), banana (3,739 individuals, 51 species, 42 genera and 19 families) and mixture of banana and coconut (3,151 individuals, 55 species, 47 genera and 19 families). Species richness ranged from 10.4-24 (mean=15.21±4.94) spiders per 0.7 m^2 mylar enclosure in irrigated rice field and its surrounding areas. In summer, irrigated rice field and their surrounding areas yielded a total of 74 taxa, 53 genera belonging to 20 families of spiders. The number of families was higher in banana and edge of the bund > coconut > irrigation canal than in irrigated rice field, *Paspalum* and other habitats.

Diversity in plant types and their landscapes is presumed to have supported the higher number of spider individuals in Candelaria, Quezon compared to that of San Juan, Batangas. It must be noted that the latter is dominated by grasses and short broad leaf plants subjected frequently to perturbations, such as animal grazing and burning. These two factors contributed to low spider diversity due to loss of suitable foundation web sites for the weavers, and low prey density in overgrazed and burned habitats for the hunting spiders. In terms of feeding guilds or mode of life following (Hatley and Macmahon, 1980) studied Philippine riceland and divided spiders into two groups i.e. the web builders and the hunters. Species-wise, there is no significant difference between the web builders (39 species) and the hunters (41 species). However, the hunters are significantly more diverse than the web builders in terms of families comprising 14 and 6 families, respectively. Among the web builders, the most diverse are families rcpresenting the comb-footed spiders, Theridiidae (12 species); long-jawed spiders, Tetragnathidae (10 species); garden spiders, Araneidae (8 species) and the dwarf spiders, Linyphiidae (7 species). On the other hand, among the hunters, the jumping spiders, Salticidae (9 species); crab spiders, Thomisidae (8 species) and the wolf spiders, Lycosidae (6 species) represent the most diverse group.

B. Population Dynamics and Movements

Many researchers have provided descriptions of spider species abundance or composition in variety of agroecosystems (Wisnieswka, 1997). The peak population density of spiders coincides with an increase of insect pests (Kiritani *et al.*, 1972). An increase in the spider population depends on prey availability and therefore the density of prey becomes higher, spiders are expected to increase proportionally to some extent. Habitat complexity is directly related to spider abundance and many spiders live strictly in specific environments related to the kind of vegetation present (Foelix, 1982). According to Rosenzweig (1995), the main parameters of richness and abundance of spiders can be affected by diverse factors such as: seasonality, spatial heterogeneity, competition, predation, habitat type, environmental stability and productivity. The rice culture is relatively simple, with few substrates for web building and hunting, although its complexity increases with time as the plants grow. In the rice culture, abundance increases up to the end, when plants reach their full size; after the harvest, abundance decreases rapidly. Also, on both occasions where no spiders were found in a transect this happened in rice areas shortly after sowing. Bambaradeniya and Edirisinghe (2001) demonstrated that abundance along rice development can be very variable; Corseuil *et al.* (1994b) recorded lower abundances both at the beginning and the end of their sampling period. Ambalagan and Narayanasamy (1999) argue that both spider abundance and richness is related to the different stages of rice growth. Such pattern of dependence on plant substrate is thus probably universal Fig. 1 (Lee and Kim, 2001).

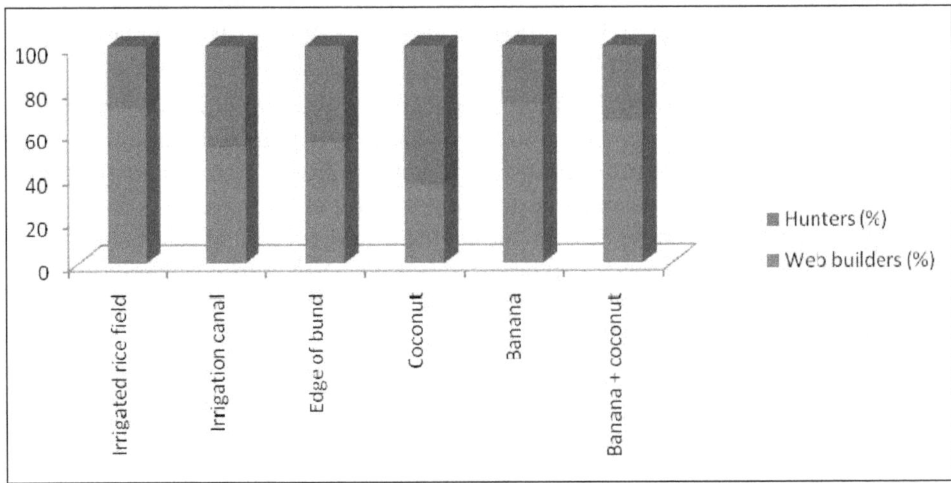

Fig. 1: Spider communities with respect to surrounding habitat *(Source-Lee and Kim, 2001)*

Spiders are always associated to rice agroecosystem due to the abundance and richness of prey in this habitat. In irrigated rice field, spiders are visible on the rice canopy and above water environments during or immediately after transplanting in both wet season (September to November) and dry season (February to April). In wet season, population peak occurs in October and in March for dry season. Both coinciding to the maximum tillering stage of the rice plant. During harvest and fallow periods, some spiders still remain in the rice fields but in low densities. Generally, these are hard to find because the spiders hide themselves underneath soil cracks and root systems of the stubbles. Many spiders migrate to the surrounding areas after harvest. These nearby habitats around rice fields serve as refuge areas for these predators. Here, they fulfill their nutritional requirements for survival and reproduction as these areas are house to numerous preys, such as the collembolans, fruit flies and other dipterans, moths and butterflies, leafhoppers and planthoppers, etc. The peaks of abundance of spiders in non-rice habitats are in August, September and December. The lowest population was in November and during summer months of April to June when grass cover is minimal, stunted and almost burned due to drought (Fig. 2).

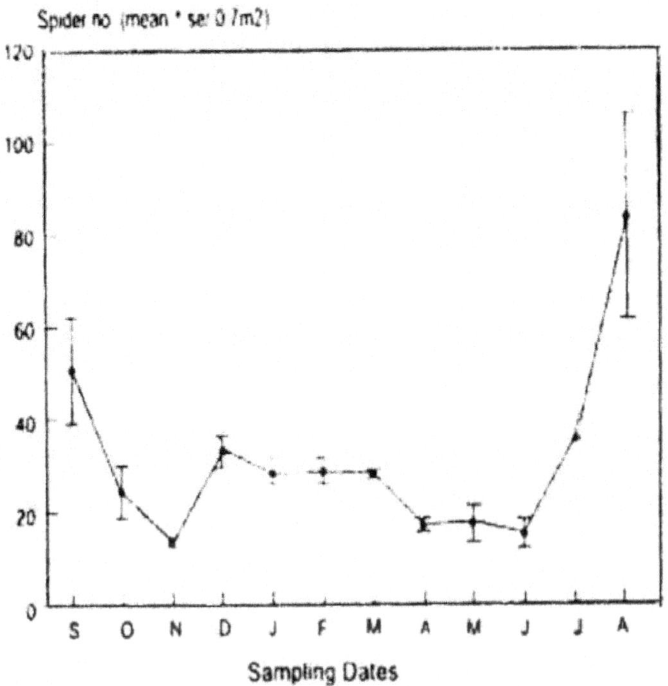

Fig. 2: Seasonal Population Trends of Orb-Weavers and Hunting spiders *(Source:-Barrion, 2001)*

In terms of movements, spiders switch between habitats for shelter and search of their nutritional requirements. Transparent mylar films thinly coated with tangle foot on both sides are placed vertically around rice field or specific habitat as enclosure strongly demonstrated that during land preparation, spiders emigrate and immigrate. Fallow fields when plowed increase spider migration into newly transplanted rice fields. One day after transplanting, spider density in the rice field increased nine folds compared to rice field with surrounding unplowed fallows. The increase in spider density is extended further at 20 and 40 days after transplantation with 6.7 and 5.4 folds increase, respectively. These movements of spiders that comprise 78-89% of the wolf spiders, *Pardosa pseudoannulata* (Boesen berg and Strand) and *P. irriensis* Barrion and Litsinger and the dwarf spider, *Atypena formosana* (Oi) regulated leafhopper and planthopper densities to 18.2±6.7/ m^2 at 20 days after transplantation and 15.4±4.3/m^2 at 40 days after transplantation. This result can be one management option in spider manipulation to combat insect pest problems in the rice fields.

C. Prey and Predation Rates

Mesh size of the web of spiders was found to be related with the prey size. The mesh size was larger in the webs of orb-weaving spider as they have larger prey size (Uetz *et al.*, 1978). This is also supported by several other researchers. The difference in the prey size may be related to the quality of the web.

A relationship between mesh size and prey length has also been studied by Uetz *et al.* (1978). He observed that a lower mesh size targets the prey items with smaller body length that otherwise may pass through a web with larger mesh size. According to Eberhard (1990), a narrow mesh size may facilitate the capture of larger prey, as more threads are in contact with the prey. However, more spiral turns decrease the distance between two turns and reflect more light thus increasing the visibility of web to prey (Craig, 1986) and reducing the prey capturing efficiency of web. Mesh size may therefore indicate a compromise between prey retention and web visibility. A larger capture area results in a higher prey interception rate (Craig, 1986) and by increasing the distance between sticky spirals spiders may enlarge overall capture area without greater energy expenditure. Accordingly, food deprived spiders commonly increase web area to enhance prey encounter (Sherman, 1994).

Spiders have a wide variety of prey. Barrion (1981) documented that in rice agroecosystems of Philippines, the prey records of 48 taxa of Riceland spiders in their natural habitat while in farmer's fields in the different cropping systems sites throughout the Philippines from 1977 to 1981. Among the spider group observed are the 10 most common taxa that are key biological control agents against rice insect pests (Barrion and Litsinger, 1984). Most of the spider preys are soft-bodied phytophage insects viz. chironomids, collembolans, leaffolders, leafhoppers, planthoppers, ricebugs, stemborers, and whorl maggots in the rice ecosystems. The entire prey spectrum consisted of 198 species belonging to 91 families in 14 orders, all belonging to Class Insecta. Among these 98 species (48.9%) are rice insect pests. A small portion of preys caught in the webs are beneficial species (Barrion, 2000). Araneophagy was not observed among the five web building spiders. Some preliminary observations are available for the spiders with hunting behaviour but a more rigorous documentation is still needed at this point to make conclusive statements. For instance, Li *et al.* (1997) reported that flies are the prey component for the jumping spiders. The 120 records of spitting spider, *Scytodes* (Scytodidae) feeding were observed in the field. Insects accounted for 19% (n=23) of all the preys and the other 81% (n=97) were spiders. Among the scytodid preys are leafhoppers, planthoppers, moths, lacewing, crickets and preying mantis. Knowledge of prey quality is important in the utilization of spiders in natural biological control and spider management. In rice field spiders such as *Pardosa pseudoannulata* and *Atypena formosana,* mixed diet of fruitflics, leafhoppers and planthoppers and collembola provided the nutritional requirements for their survival. These spiders may live on collembola or fruitflies alone but mixed diet is essential to optimize fecundity. Similarly, pollen feeding in rice by the orb web spider, *Argiope catenulata* (Doleschall) has been found to increase female spider fecundity.

The insect predation rates of spiders are known only to a limited number of taxa. For instance, *Pardosa* spp. consumed 4th instar nymphs of green leafhoppers (23), brown planthopper (29) and white backed planthoppers (27). This is remarkable for this predator because its "wasteful killing" behaviour inflicted high mortality to the developing sucking pests that otherwise are transformed to full adults and produce another generation of pests. Gavarra and Raros (1975) studied this phenomenon in the 3rd generation of brown planthopper nymphs using *Pardosa pseudoannulata* (Boesen berg & Strand) and reported total prevention of rice due to hopperbum injury. Mass rearing of spiders, however, is not cost effective but still time consuming. Sigsgaard and Villareal (1999) determined functional response of *A. formosana,* dwarf spiders based on the number of preys eaten per predator at different prey densities. The workers found the functional response of *A. formosana* similar to Hollings type I, II, and III functional equations with strong preference to the second instars of green leafhopper, second and third instars of the brown planthopper.

D. Seasonal Occurrence of Spiders in Rice Field

Seasonal fluctuations of spiders in rice fields in Korea have been reported to follow an "M" shaped curve (Choi and Namkung, 1976). Other surveys which began at a different time and covered a different period have reported that fluctuations in the density of spider communities followed a "W" shaped curve (Choi and Namkung, 1976). During the rice growing period in Korea, rice field spiders generally show two peaks of abundance. One is when spiders which have overwintered nearby migrate into the paddy, and the other is when the second generation of the year appears. However, annual fluctuations in number of rice field spiders showed several peaks when overwintering periods are included (Fig. 1). In rice fields of Korea, spiders start to build up their populations in the middle of July i.e. 40-45 days after transplanting (Yun, 1997).

The same pattern of spider abundance was seen in Japan, which has a similar agricultural environment (Kobayashi and Shibata, 1973). Hamamura (1969) has reported that the low density of spiders in rice fields of Japan during transplantation season. Lee *et al.* (1997) suggested that the different time taken by different spider species to immigrate from levees into paddy fields is the cause of the low initial population densities, in spite of the abundance of prey. However, Kim (1998) showed that the immigration rate of overwintering spiders from levees into paddy fields was low. He suggested that the immigration of the spiders was delayed after transplantation because the young rice plants are too small to provide a suitable habitat for spiders. He further suggested that most of the dominant spider species immigrated from nearby areas into the rice fields by ballooning (Note: "Ballooning" is the term used to describe the habit of young spiders of sailing through the air borne by silk strands on wind currents). The spiders population always shows fluctuation with crop stages and pest population. Jayakumar and Sankari (2010) studied spider population and their predatory efficiency in different rice establishment techniques. The workers recorded total of five spiders species namely *Lycosa pseudoannulata, Callitrichia formosana, Tetragnatha javanas, Argiope catenulata* and unidentified Plexippus species. They reported that except *T. javanas,* all the spiders were observed throughout the study. *T. javanas* was higher in early growth stage and *A. catenulate* was predominant during the later stage of the crop.

The web building and plant wandering spiders rely on vegetation for some part of their lives, either for finding food, building retreats or for web building. The structure of vegetation is therefore expected to influence the diversity of spiders found in the habitat. Many studies demonstrated that correlation exist between the structural complexity of habitat and species diversity (Uetz, 1979).

Vegetation structure seems to influence the spider composition on family level because similar families cluster within a similar habitat type. Vegetation architecture plays a major role in the species composition found within a habitat and vegetation which structurally more complex can support higher diversity. Difference in vegetation architecture during crop growth accounts for the different community structure of spiders. In addition, the difference in the seasonal abundance of spiders may be due to the variation in patterns of activity of individual spiders and phenology of total spider community (Corey *et al.*, 1998). It was also reported that similar species of spiders are present at specific stage of crop growth. Thus, vegetation structure may be a more important determinant that their seasonal variation alone.

References

Ambalagan, G. and Narayansamy, P. (1999). Population fluctuation of spiders in the rice ecosystem of Tamil Nadu *Entomon Trivadrum* **24 (1)**: 91-95.

Bandaradeniya, C. B. and Edirisinghe, J.P. (2001). The ecological role of spiders in the rice fields of Sri Lanka *Biodiversity* **2:** 3-10.

Barrion, A.T. (198l). The spider fauna or Philippine dryland and wetland rice agroecosystcms. MS thesis, Ocpanment of Entomology, UP Los Banos, Laguna.

Barrion, A. T. (2000). Vcnical stratification and prey spectrum of web building spiders in irrigated nee fields and selected non-rice habitats in three Southern Tagalog provinces in the Philippines *Asia life Sciences* **9(1):** 67-101.

Barrion, A. T. and Litsinger, J. A. (1984). The spider fauna of Philippine rice ugroecosystcms. II. Wetland **6(1):**11 -37.

Barron, A. T. (2001). Spiders natural biological control agent against insect pests in philippine rice field. *Trans. Natt.Acad. Sci & Tech. Phillippines* **23:** 121-130

Choi, S, S, and Namkung, J, (1976), Survey on the Spiders of the Rice Paddy Field (I) *Korean Journal of Plant Protection* **15(2):** 89-93.

Corey, D. T., Stout, I. J. and Edwards, G. B. (1998). Ground surface spider fauna in Florida sandhill communities *Journal of Arachnology* **26:** 303-16.

Corseuil, E., Paula, M. C. and Brescovit, A. D. (1994). Aranhas associadas a uma lavoura de arroz irrigado no município de Itaqui, Rio Grande do Sul *Biociências* **2(2):** 49-56.

Craig, C. L. (1986). Orb-web visibility: The influence of insect flight behaviour and visual physiology of the evolution of web designs within the Araneoidea. *Animal Behaviour* **34:** 54–68.

Determinants of web spider species diversity: vegetation structural diversity vs. prey availability *Oecologia* **62:** 299-304.

Eberhard, W. G. (1990). Function and Phylogeny of spider webs Annual Review of Ecology, Evolution and Systematics **21:** 341–72.

Foelix, R. F. (1982). Biology of spiders. Cambridge, Harvard University Press. 306 p

Hamamura, T. (1969). Seasonal fluctuation of spider populations in paddy fields. *Acta Arachnology* **22 (2):** 40-50.

Hatley, J., Ross, D. W. and Moldeken, A. R. (1980). Spider community organization: Seasonal variation and the role of vegetation architecture. Environmental Entomology **9**: 632-39.

Jayakumar, S. and Sankari, A. (2010). Spider population and their predatory efficiency in different rice establishment techniques in Aduthurai, Tamil Nadu. *Journal of Biopesticides* **3(1):** 020 - 027

Jayakumar, S. and Sankari, A. (2010). Spider population and their predatory efficiency in different rice establishment techniques in Aduthurai, Tamil Nadu. *Journal of Biopesticides* **3(1)**: 020 – 027.

Kim, H. S. (1992). Suppressive Effects of Wolf Spider, Pirata subpiraticus (Araneae: Lycosidae) on the Population Density of Brown Planthopper (Nilaparvata lugens Stål). Unpublished Ph.D. Thesis, Dongkuk Univ, Seoul, Korea. 69pp.

Kim, S. T. (1998). Studies on the Ecological Characteristics of the Spider Community at Paddy Field and Utilization of the Pirata subpiraticus (Araneae: Lycosidae) for Control of Nilaparvata lugens (Homoptera: Delphacidae). Unpublished Ph.D. Thesis, Konkuk Univ., Seoul, Korea, 90 pp.

Kiritani, K., Hokyo, N., Sasaba, T. and Nakasuji, F. (1970). Studies on population dynamics of green rice leafhopper, Nephotettix cincticeps Uhler: Regulatory mechanism of the population density *Researches on Population Ecology* **12:** 137–53.

Kiritani, K., Kawahara, S., Sasaba, T. and Nakasuji, F. (1972). Quantitative evaluation of predation by spiders on the green rice leaf hopper, Nephotettix cincticeps Uhler, by a sight count method *Researches on Population Ecology* **13:** 187-200.

Kobayashi, S. and H, Shibata. (1973). Seasonal changes in population density of spiders in paddy fields, with reference to the ecological control of the rice insect pests. *Japanese Journal of Applied Entomology and Zoology* **17(4):** 193-202.

Lee, J. H. and Kim, S. T. (2001). Use of spiders as natural enemies to control rice pest in Korea. Available in the World Wide Web at: . [Last access: [16 apr 2005].

Lee, J. H., Kim, K. H. and Lim, U. T. (1997). Arthropod community in small rice fields associated with different planting methods in Suwon and Icheon Korean Journal of Applied Entomology **36(1)**: 55-66.

Li, D., Jackson, R. R. and Barrion, A. (1997). Prey preferences of Portia labiata, P. africana, and P. shultzi, araneophagic jumping spiders (Araneae: Salticidae) from the Philippines, Sri Lanka, Kenya and Uganda New Zealand Journal of Zoology **24:** 333- 49.

Rodrigues, N. L., Milton, D. S. and Richardo, O. T. (2009). Spider diversity in a rice agroecosystem and adjacent areas in southern Brazil. *Revista Colombiana de Entomología* **35 (1)**: 89-97.

Rosenzweg, M. L. (1995). Species diversity in space and time. Cambridge, Cambridge University Press. Pp-436.

Scheidler, M. (1990). Influence of habitat structure and vegetation architecture on spiders *Zoology Anzeiger* **225:** 333-40.

Settle, W. H., Ariawan, H., Astuti, E. T., Cahyana, W., Hakim, A. L., Hindayana D., Lestari, A. S. & Sartanto, P. (1996). Managing tropical rice pests through conservation of generalist natural enemies and alternative prey *Ecology* **77:** 1975– 88.

Sherman, P. M. (1994). The orb-web: an energetic and behavioral estimator of a spider's dynamic foraging and reproductive strategies *Animal Behavior* **48:** 19-34.

Sigsgaard, L. and Villareal. S. (1999). Predation rates of Atypena formosana Oi) onbrown planthopperand greenleafhopper. *International Rice Re search Notes*. **24(3)**: 18

Tahir, H. M. and Butt, A. (2009). Predatory potential of three hunting spiders inhabiting the rice ecosystems Journal of Pest Science **82:** 217-25.

Uetz, G. W. (1979). The influence of variation in litter habitats on spider communities *Oecologia* **40:** 29-42.

Uetz, G. W. (1990). Prey selection in web-building spiders and evolution of prey defenses. In: Evans DL, Schmidt OJ (eds) Insect defense. Adaptive mechanisms and strategies of prey and predators. University of New Yark Press, Albeny, pp 93–128

Uetz, G. W. (1991). Habitat structure and spider foraging. Habitat Structure: The Physical Arrangement of Objects in Space (ed. by S. Bell, E. McCoy and H. Mushinsky), pp. 325–348. Chapman & Hall, London, UK.

Uetz, G. W., Johnson, A. D. & Schemske, D. W. (1978). Web placement, web structure, and prey capture in orb-weaving spiders *Bulletin of British Arachnological Society* **4:** 141-48

Wisniewska, J. and Prokopy. R. J. (1997). Pesticide effect on faunal composition, abundance, and body length of spiders (Araneae) in apple orchards *Environmental Entomology* **26:** 763-76.

Yoon, J. C. (1997). Arthropod Community Structure and Changing Patterns in Rice Ecosystems of Korea. Unpublished Ph.D. Thesis, Seoul National University, Seoul, Korea, 105 pp.

Chapter 7

Conservation of Natural Enemies

Maninderjit Singh, Amandeep Singh and Randeep Singh

(a) Habitat Management

Promoting generalist predators in agriculture *via* habitat manipulation has gained much interest in biocontrol research during the recent past. This strategy helps in managing the insect pests by enhancing natural enemies without adversely affecting the main crops and also reduces the pesticide load.

Unlike many insect groups, spiders do not have strong host plant associations, although some do select plants for their physiognomy. In some species e.g. *Pardosa lugubris* (Walckenaer), there may be differences in microhabitat choice between adults and juveniles in order to facilitate prey capture and avoid predation. Various measures of habitat structure have been used to explain the diversity or abundance of spiders: vegetation tip height diversity (Greenstone, 1984); vegetation height; vegetation density (Hatley and MacMahon, 1980); and proportion of bare ground.

There may be different reasons why such variables correlate with spider diversity and abundance, but it is thought that web spinners have the strongest relationship with physical structures (Uetz, 1991). However, even for nonweb spinners, habitat structure can be important, e.g. *Dolomedes plantarius* (Clerck) selects one species of sedge (*Cladium mariscus* (L.)) for oviposition sites. In addition, it was found that jumping spiders (Salticidae) used horizontal and vertical structures in different ways and preferred widely spaced to dense structures. It is not only vegetation architecture that is important but also its state. Duffey (1962) found that the proportion of living plants determine the number and type of webs that are constructed, and noted that the non-webspinning clubionids used only living plants in which to make retreats. Furthermore, surfaces of plants are used during courtship to conduct vibrations (Uetz, 1991), tips of vegetation are platforms for ballooning spiders and tussocks of vegetation are good overwintering sites. Although vegetation plays a vital role, abiotic structures (such as water and stones) and man-made structures are also used. For example, two linyphiid species, *Leptorhoptrum robustum* (Westring) and *Erigone longipalpis* (Sundevall) were able to live permanently under ®lter beds in sewerage pits. Some spiders such as *Ariadna masculine* Lawrence (Segestriidae) use specific types of stone placed outside burrows to extend the foraging range by providing a larger area of substrate suitable for attaching its web.

Plate-31: Straw bundles from sorghum to rice field

Rice

In rice fields, the spiders which are the most abundant predators present, tend to take refuge on field bunds. Studies have shown that addition of straw bundles placed in sorghum fields to attract spiders and then transferred to rice fields significantly reduced stem borer and leaf folder populations. In rice ecosystem where there is a plethora of natural enemies, CBC would play a major role in reducing pests through enhanced natural control. Some techniques that can be successful in rice field are as follows:

- Placing straw bundles on bunds to provide shelter for spiders.
- Increasing flowering weed/crop flora on the bunds to provide nectar, pollen and alternate host
- Application of organic manures to increase population of benthic flora that will help early colonization of spiders and other generalist predators.
- Alleyways to increase predator action in BPH prone areas
- Placing bird perches at the field edge to increase predation of caterpillars and rats by birds

Field trials were conducted during 2006–07 at Atterna village (Sonipat–Haryana) on habitat management through artificial structures and border cropping with other suitable crops. The experiments were conducted with Pusa Sugandh 4 (Pusa 1121) as the main crop with 5 treatments i.e. straw bundles+maize (T1), straw bundles+sunhemp (T2), straw bundles+*Sesbania* (T3), only straw bundles (T4) and control (T5) without any additional structure or border crop. Each plot was of 0.5 acre in size. The straw bundles were prepared with wheat straw placed in plastic nets. Each bundle had 3 feet length and 10 inches diameter. Both the ends of these bundles were tied with plastic rope. These bundles were placed in sorghum fields for charging with spiders and other natural enemies. After 15 days of charging, these bundles were fixed vertically with bamboo sticks @20 bundles/ha in paddy field having 20 days transplanted seedlings in such a way that the lower portion of the bundle remained 6 inches above the water level. Control plot involved two insecticide sprays, in August and the other in September. The observations on the infestation and population of natural enemies especially spiders were recorded from twenty hills selected randomly in each plot after 10 days interval. In all the treatments (T1 - T4), the population of leaf folder and yellow stem borer remained lower than the control (T5). The observation recorded on charging of the straw bundles for 15 days in sorghum fields indicated that each bundle contained 8–10 spider egg masses, 30–40 spider adults, 500–600 spiderlings and 20–30 earwigs. The results on the population dynamics of spiders indicated that the adult spiders on paddy plants remained highest in the treatment having only straw bundle (T4) followed by straw bundles+sun hemp (T2), straw bundles+*Sesbania* (T3) and straw bundles+maize (T1). Control (T5) indicated lowest population of spiders throughout the season. In September, maximum spiders confined to the lower part of plant (upto 10 inches above the ground) as compared to middle (10–20 inches above the ground) and upper part (20 inches above the ground). The population of spiders on ground remained highest in treatment T_4 followed by T_3, T_1 and T_2. Most of spiders on ground belonged to two families i.e. Lycosidae (Wolf spiders) and Salticidae (Jumping spiders). The results of present study indicated that straw bundles along with border crops had played an important role in conservation of spiders in rice which has helped to contain the insect pests and reduce the pesticide load. Harvest of the crop indicated highest paddy yield in T_2 followed by T_3, T_4, T_1 and lowest in T_5.

(b) Innovative Methods and Techniques

One of the major constraints in the mass production of natural enemies is the non-availability of low cost and efficient mechanical devices particularly under field conditions. In the absence of such devices and techniques, lot of manual labour is required for the mass production and maintenance of natural enemies. Further, the availability of these biocontrol agents is a serious constraint in promotion of IPM. A project was initiated to design, develop and standardize the low-cost methods and techniques for the mass multiplication of natural enemies of insect pests directly under field conditions.

The most common rearing method consist of confining the spider in a container of some sort, opening the container from time to time to add food and water. The containers consist of: a) glass jars and tumblers covered by wire screen (Bonnet, 1930), cheesecloth (Burger, 1937), mosquito netting (Crane, 1948) and cotton cloth (Hollis and Branson, 1964); jars with holes in the metal lids (Rovner, 1968); and jars stoppered with either polyurethane foam or cotton wrapped in cheesecloth (Kaston, 1972); b) glass vials and tubes covered with corks (Branch, 1942; Cooke, 1962), cotton (Branch, 1942; Nakamura, 1968), and polyurethane foam (Kaston, 1970); c) finger bowls covered by glass, plastic boxes (Cooke, 1962; Miyashita, 1968); e) cardboard ice cream containers covered by petri dishes or glass plates (Whitcomb and Eason, 1965); f) for orb weavers (Araneidae), aluminum frames with removable glass doors (Witt, 1971). Each spider was reared in a separate container to prevent the cannibalism. Usually live insects such as house flies (*Musca domestica*) and Drosophila, have been used for food most often, although some work has been done with artificial diets (Peck and Whitcomb, 1968). Feeding methods frequently include anesthetizing the flies for sorting and transferring, although Brown (1946) described a useful method for transferring individual Drosophila without anesthesia.

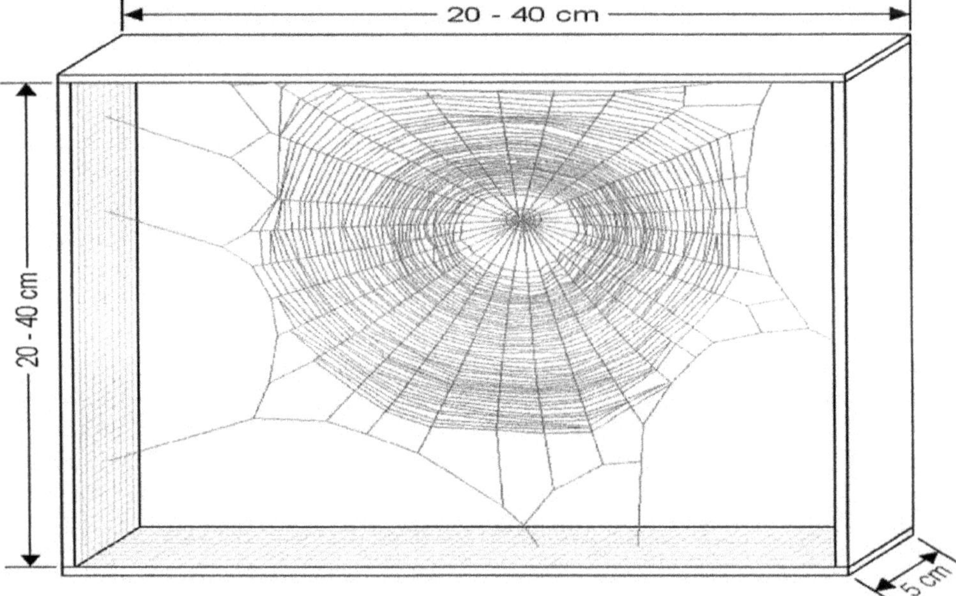

Plate-32: Frame to keep orb-web spiders made out of four Perspex strips, glued together at the corners (not to scale). On the inside of the frame, blackened crack-seal tape has been applied to facilitate the spider's grip. Similar frames with dimensions accordingly adapted can be used for other web-building spiders.

Depending on the species, water was also provided for drinking (Parry, 1954) and/or for humidity may be important. Some provisions include placing the following items inside the spider 's container: a glass dish filled with water (Bonnet, 1930), a wet cork (Brown, 1946), wet cotton (Hollis and Branson, 1964), wet cotton covered with filter paper (Nakamura, 1968), a wet cotton roll (dental wad) (Crane, 1948), small vials and tubes filled with water and plugged with cotton (Miyashita,

1968 ; Rovner, 1968), wet plaster of Paris (calcium sulfate), and wet plaster of Paris mixed with sand (Cooke, 1962). Another method used by Bonnet (1930) was to fill the bottom of jars with water. A ring of cork floating in the water, or a stick of wood dropped inside of the jar that served as a substrate for the spider. Parry (1954) kept spiders in flower pots containing soil and covered with glass. Soil moisture was maintained by keeping the trays of water in flower pots. Another study by Burger (1937) used cheesecloth covered glass tumblers. Food and water were introduced through a thistle tube making a hole in the cheesecloth. Cooke (1962) covered glass tubes with corks in which there were two holes contained a smaller glass tube. One was filled with water and covered with cotton at both ends. The other tube, used for ventilation and introduction of prey, was covered at one end with cotton.

Enclosures (frames) to keep spiders

A variety of frames have been used to study spiders and their webs. Witt (1971) used elaborate metal cages for the study of spiders. He suggested simpler frames entirely made out of Perspex. The frame's size should correspond to the web size. For this initial field measurements are necessary. To study small to medium sized orb-web spiders (e.g., *Zilla diodia*, juvenile *Araneus diadematus* or *Larinioides sclopetarius*, frames consisted of four pieces of transparent Perspex having dimensions of 5 × 30 × 0.3mm (wide × long × thick) glued together with industrial strength glue at the corners (Fig. 2). Large orb-web spiders (e.g., adult *Argiope* sp.) require frames made out of 50 cm long Perspex pieces and adult *Nephila* sp. require even larger frames. Similar frames, but kept horizontally, can be used for sheet-web spiders (Bartels, 1929). Spiders building three-dimensional webs (e.g., linyphiid and theridiid spiders) require cube shaped frames. For some species of spiders, it can be advantageous to build the frames higher than wide.

To facilitate the spider's grip inside the frame, apply net-like crack-seal tape, painted black to reduce unwanted reflections when later taking pictures of the web. To allow unobstructed examination of the web, the spiders must be kept in frames where two opposite sides can be removed. To separate the frames, place thin (0.5 mm), large (a few cm larger than the frames), transparent and somewhat flexible PVC sheets between them. These sheets are smeared with vaseline to deter spiders from attached threads. Alternatively, windowpanes, which are kept very clean, can be used. In addition, puffy foam can be put along the edge of the frame, encouraging spiders to attach to that foam rather than to the glass. The frames are put on a shelve like books with the thin sheets of PVC placed between them (Fig. 32). So that when spider has built a web, the frame can be easily taken from the shelf and placed in front of a shadow box for examination. Remove the frame carefully, because spider usually stays in its web (or retreat). Some freshly caught spiders are likely to leap off the web or leave the hub of the web when frame is removed for the first time, but will mostly become habituated to this after few days. For short term or preliminary experiments, there are many alternatives to the durable Perspex frames, such as rigid cardboard, wooden frames or slices of round plastic buckets with cling wrap to prevent spiders from escaping. Spiders building webs on rather than between supports can be offered an artificial, standardized structure to build their web on (Blackledge & Wenzel, 2001), which is then placed inside a larger container with clean, smooth sides. For short-term storage of smaller spiders, upturned plastic cups are used from which the bottom has been removed and replaced with a fine mesh. Smaller spiders will build small webs in these cups, which can be misted from the top (with a spray bottle) without lifting the cup. Alternatively, only a small hole is cut into the bottom of the cup and corked with a cotton plug or a tampon piece and moisture is provided by placing moist cotton plug. This type of rearing is also used for experimental studies such as web-building behaviour, larger spiders can be maintained in cups to temporarily prevent them from building webs (Reed *et al.*, 1970; Herberstein *et al.*, 2000).

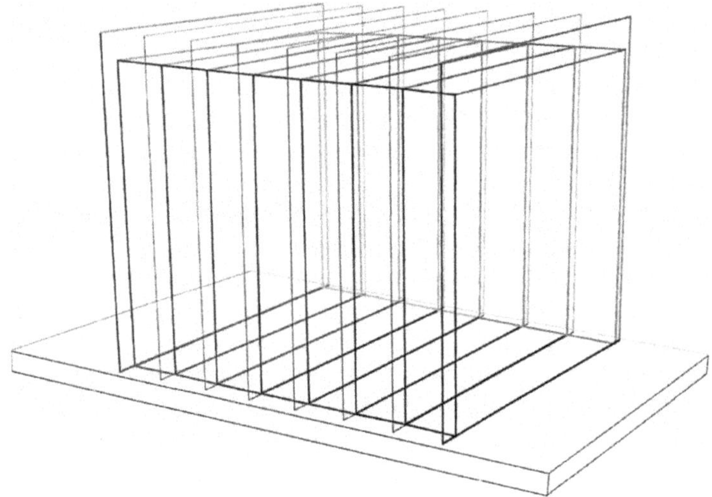

Plate-33: Several frames put side to side on a shelf. Thin PVC sheets placed between adjacent frames prevent the spiders from moving between frames. The thin sheets are smeared with Vaseline to prevent the spiders from attaching threads. The first and the last sheet are thicker, more stable ones, held up with bookends.

Feeding and watering

Spiders can feed on almost all kinds of insects, and most web-building spiders will attack and overwhelm insects trapped in their webs. The size of prey insect vary from small to large or even larger than their own size. The Drosophila flies are easily reared in laboratory are used as food for spiders. When rearing Drosophila in bottles with sponge stoppers, spiders can be fed by trapping single flies between the flange of the stopper and the bottle, from where they can be introduced into the web using forceps. Spiders without webs are trickier to feed. Some non web building spiders accept live prey held near their mouth with forceps and prefer buzzing flies as compared to kicking crickets. Sprinkling water over the offered insect or breaking the insect's cuticle a bit by snipping a cercus or antenna to release a drop of hemolymph can induce spiders to feed when the liquid touches the spider's chelicerae. Most of spiders feed once per week. However, non web building spiders feed twice per week. Feeding of spiders must be sufficient so that they do not loose substantial amounts of weight, which leads to shrinkage of abdomens. Whereas some spiders can be kept on minimal diet for prolonged time e.g., *Araneus diadematus, Zygiella x-notata,* other species seem to feeble when they are not given enough food required for growthto growth e.g., *Argiope bruennichi.* Natural prey capture rates may provide helpful starting points when designing feeding regimes in the laboratory. It is important to either feed the spiders with different insects or to feed the prey insects with high quality food (i.e. supplemented with proteins, vitamin-enriched cereal or pet food), as the spiders may otherwise experience deficiencies (Uetz *et al.,* 1992; Mayntz and Toft, 2001). The spiders kept in relatively dry air in most buildings are vulnerable to desiccation. Thus regular misting with a water sprayer or placing a moist sponge at the bottom of the frame is necessary.

Plate-34: Laboratory rearing of drosophila for spiders

Rearing spiderlings

This is difficult procedure and often result in high levels of mortality. To rear spiderlings, place the freshly hatched cocoon into a container with support for the webs building such as wood-wool, and add cultures of Drosophila or Collembola as food (Dinter, 2004). Initially, some spiderlings will show **cannabalism,** but more established ones will construct webs and capture prey. Do not separate the spiderlings too early as this can lead to 100% mortality.

Encouraging web building

Spiders vary greatly in their tedency to build a web in the laboratory. The most popular orb-web species include *Araneus* sp., *Argiope* sp., *Nephila* sp. and uloborid spiders. In contrast, *Gasteracantha* sp., *Tetragnatha* sp., *Meta* sp., *Metellina* sp. and

Leucauge sp. are more hesitant to build webs. Feeding a spider or putting a live fly along with spider into cage that has not yet built a web in the laboratory (Pasquet *et al.*, 1994) can help to induce web building. The release of live fly into the frame with the spider is difficult because of strict feeding regimes, flies can be kept in a small jar with some sugar solution and covered by fly mesh. The flies will buzz and stimulate web building without being captured by the spider (Herberstein *et al.*, 2000). When spiders are exposed to natural day-night duration of light and temperature, web building frequency is increased (Witt, 1956). Web building frequency varies between species. Whereas, *Araneus diadematus, Argiope sp., Larinioides sclopetarius* and *Zygiella x-notata* generally rebuild the capture area of their web every night or every other night, while *Nephila* sp. typically rebuild only sections of the web.

Damaging webs

Some orb-web spiders hesitate to rebuild their web as long as it is intact. To induce web rebuilding, it may therefore be necessary to damage or destroy the webs. In the field, spiders generally leave the frame and the anchor threads largely intact when rebuilding the web (Carico, 1986). Thus, only the capture area should be damaged to induce rebuilding, e.g., by cutting holes into the capture area with a red-hot wire (Plate-33). Alternatively, cut the lateral anchor threads with scissors to destroy the entire orb-web. However, complete web destruction forces the spider to build the next one from scratch (Zschokke and Vollrath, 2000). In general, the spider should be allowed to ingest the old web (Peakall, 1971). Damaging a single sticky spiral segment allows to determine whether the spider rebuilds the web during the next night. Non orb-web spiders do not remove, ingest and rebuild their web as orb-web spiders do, but keep repairing and extending them (Tanaka 1989; Benjamin & Zschokke, 2003, 2004).

(c) Among Management of Ridges Between Rice Cultivation

Different factors influence the population of spiders. The number of wolf spiders hardly changed from August to September in the field where the ridges were not mowed (Fig. 3). In contrast, the number of wolf spiders increased considerably after the ridges were mowed and these spiders were observed under the cut plants in the ridges. Thus far, it is not clear whether spiders harbor under the cut plants for propagation or because these plants are a merely a suitable habitat.

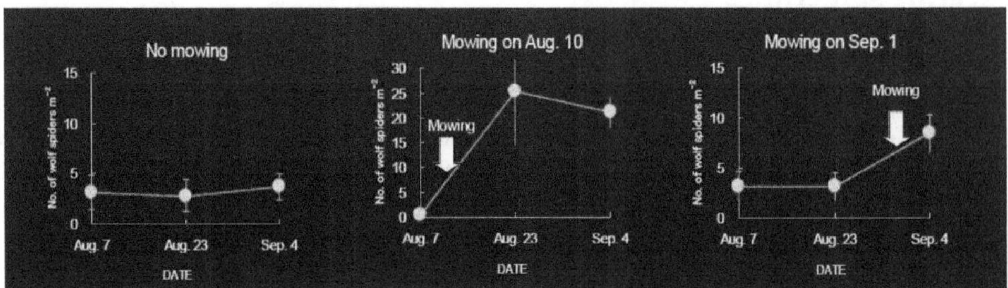

Fig. 3: Effect of the time of mowing on the number of wolf spiders in rice paddy field Error bars indicate standard deviation *(Source- Inagak et al., 2012)*

(d) Using Cover Plants (Chinese Milk Vetch) for Previous Crop in Rice Cultivation

Chinese milk vetch (*Astragalus sinicus* L.) cultivation is traditionally used as green manure and living mulchin rice paddies. The number of wolf spiders in rice paddy fields cultivated Chinese milk vetch was larger than that in fields growing water foxtall or fields with no vegetation (Fig. 4). We think that Chinese milk vetch cultivation effects an increase in the population of wolf spiders.

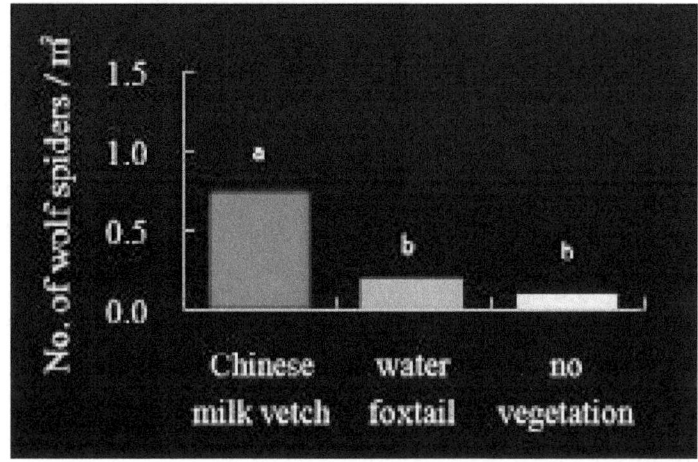

Fig. 4: The number of wolf spiders in rice paddy fields before rice transplanting (Tukey's at p=0.05) *(Source- Inagak et al., 2012)*

From the above studies, it is concluded that management practices such as moving ridges and Chinese milk vetch cultivation before rice transplanting effects an increase in the population of wolf spiders in rice paddy fields which in turn will control of rice insect pests.

Plate-35: Chinese milk verch (left) and water foxtail (right)

(e) Provision of Refuge

Refuges have long since been advocated for early colonization and conservation of generalist predators. Refuge provides the natural enemies with nectar sources, pollen and alternate prey in addition to providing shelter habitat. Predator refugia are provided in the form of intercrops, cover crops, field margins, hedgerows, fencerows, windbreaks etc.

The aim of shelter habitats is to provide suitable conditions for overwintering, aestivation, reproduction and a general safety from anthropogenic disturbances in farmland. They could be a source of alternate prey and also add to general biodiversity of crop lands. Field margin refuges around arable fields have been reported to enhance the within-field activity and density of polyphagous predators.

References

Bartels, M, (1929). Sinnesphysiologische und psychologische Untersuchungen an der Trichterspinne Agelena labyrinthica (Cl.) *Zeitschrift fu¨r vergleichende Physiologie* **10:** 527–93.

Benjamin, S. P. and Zschokke, S. (2003). Webs of theridiid spiders: construction, structure and evolution *Biological Journal of the Linnean Society* **78:** 293–305.

Benjamin, S. P. and Zschokke, S. (2004). Homology, behaviour and spider webs: web construction behaviour of Linyphia hortensis and L. triangularis (Araneae: Linyphiidae) and its evolutionary significance *Journal of Evolutionary Biology* **17:** 120–30

Blackledge, T. A. and Wenzel, J. W. (2001). State-determinate foraging decisions and web architecture in the spider Dictyna volucripes (Araneae Dictynidae) *Ethology, Ecology and Evolution* **13:** 105–13.

Bonnet, P. (1930). La mue, 1'autotomie et la regeneration chez les Araignees, avec une etude de s Dolomedes d'Europe. **59** :237-700.

Branch, J. H. (1942). Notes on California Spiders. 1 . On the culture of the spider Teutana grossa Koch *Bulletin of the Southern California Academy of Science* **41:** 138-39.

Brown, H. P. (1946). A technique for rearing spiders. Turtox News **24** :76.

Burger, E. (1937). Culture of Lathrodectus mactans, the black widow spider pp. 244-245. In Needham, J. G., P. S. Galtsoff, F. E. Lutz, and P. S. Welch, Culture Methods for Invertebrate Animals. Comstock, New York.

Carico, J. E. (1986). Web removal patterns in orbweaving spiders. Pp. 306-318. In Spiders: Webs, Behavior, and Evolution. (W.A. Shear, ed.). Stanford University Press, Stanford.

Cooke, J. A. L. (1962). Keeping spiders in captivity. Bull. Flatford Mill Spider Group 13 :4-5

Crane, J. (1948). Comparative biology of salticid spiders at Rancho Grande, Venezuela. Part II. Methods of collection, culture, observation, and experiment. Zoologica 33 :139-145.

Dinter, A, (2004), A mass rearing method for the linyphiid spider species Erigone atra (Blackwall) (Araneae: Linyphiidae) *Journal of Applied Entomology* **128:** 200–203.

Duffey, E. (1962) A population study of spiders in limestone grassland. Description of study area, sampling methods and population characteristics. *J. Anim. Ecol.* **31**: 571-99

Greenstone, M. J. (1984). Determinants of web spider species diversity: vegetation structural diversity vs. prey availability *Oecologia* **62:** 299-304.

Herberstein, M. E., Gaskett, A.C., Glencross, D., Hart, S., Jaensch, S. and Elgar, M. A. (2000). Does the presence of potential prey affect web design in Argiope keyserlingi (Araneae, Araneidae) *Journal of Arachnology* **28:** 346–50.

Hollis, J. H. and Branson, B. A. (1964). Laboratory observations on the behavior of the salticid spide r Phidippus audax (Hentz) *Transaction of the Kansas Academy of Science* **67:**131-48.

Inagaki, H., Matsuno, K., Ichihara, M, Saiki, C., Yamaguchi, S., Mizumoto, S. and Yamashita, M. (2012). Habitat management to conserve wolf spiders, natural enemies of insect pests, in rice paddies in Japan. *7TH Symposium IPM* 90.

Kaston, B.J. (1970). Comparative biology of American black widow spiders. Transaction of the San Diego Society of Natural History **16:** 33-82.

Kaston, B. J. (1972). How to know the spiders. W. C. Brown, Dubuque. 289 p

Mayntz, D. and Toft, S. (2001). Nutrient composition of the prey's diet affects growth and survivorship of a generalist predator *Oecologia* **127:** 207-213.

Miyashita, K. (1968). Changes of the daily food consumption during adult stage of Lycosa pseudoannulata Boes. et Str. *Applied Entomology and Zoology* **3:** 203-204.

Nakamura, K. (1968). The ingestion in wolf spiders. In: Capacity of gut of Lycosa pseudoannulata. Res. Popul. Ecol. 10 :45-53. Parry, D. A. 1954. On the drinking of soil capillary water by spiders. *Journal of Experimental Biology* **31:** 218-27.

Parry, D. A. (1954). On the drinking of soil capillary water by spiders *Journal of Experimental Biology* **31:** 218-22.

Pasquet, A., Ridwan, A. and Leborgne, R. (1994). Presence of potential prey affects web-building in an orb-weaving spider Zygiella x-notata *Animal Behaviour* 47:477–480.

Peakall, D. B. (1971). Conservation of web proteins in the spider, Araneus diadematus. *Journal of Experimental Zoology* **176:** 257–64.

Peck, W. B. and Whitcomb, W. H. (1968). Feeding spiders an artificial diet. Entomol. News **79:** 233-236.

Reed, C. F., Witt, P. N., Scarboro, M. B. and Peakall, D. B. (1970). Experience and the orb web. *Developmental Psychobiology* **3:** 251-65.

Rovner, J. S. (1968). An analysis of display in the lycosid spider Lycosa rabida Walckenaer. *Animal Behavior* **16:** 358-36.

Tanaka, K. (1989). Energetic cost of web construction and its effect on web relocation in the webbuilding spider Agelena limbata *Oecologia* **81:** 459–65

Uetz, G. W. (1991). Habitat structure and spider foraging. Pp. 325-348, In Habitat structure, the physical arrangement of objects in space (S.S. Bell, E.D. McCoy & H.R. Mushinsky, eds.). Chapman & Hall, New York.

Uetz, G. W., Bischoff, J. and Raver, J. (1992). Survivorship of wolf spiders (Lycosidae) reared on different diets *Journal of Arachnology* **20:** 207– 11.

Whitcomb, W. H. and Eason, R. (1965). The rearing of wolf and lynx spiders (families Lycosidae and Oxyopidae: Araneida) *Journal of Arkansas Academy of Science* **19:** 21-27.

Witt, P. N. (1956). Die Wirkung von Substanzen auf den Netzbau der Spinne als biologischer Test. Springer, Berlin. 79 pp

Witt, P. N. (1971). Instructions for working with web-building spiders in the laboratory *Bioscience* **21:** 23-2

Zschokke, S. and Vollrath, F. (2000). Planarity and size of orb-webs built by Araneus diadematus (Araneae: Araneidae) under natural and experimental conditions. Ekolo´gia 19(Supplement 3):307–318.

Chapter 8

Spiders as Detrivores

Amandeep Singh, Abhinay Thakur, Kapila Mahajan and Mandeep Kaur

Spiders can also exert significant top-down effects, meaning that plant damage by insect herbivores is lower when spiders are present than when they are absent. Encouraging hunting spiders by the addition of mulch, which provides shelter and humidity, resulted in a significant decrease in plant damage in vegetable gardens (Riechert and Bishop, 1990).

The research on the roles of spiders in agroecosystems has focused primarily on the extent to which these predators suppress densities of grazing herbivores. Spiders also belong to decomposition food webs of agroecosystems and feed on saprophages, microphytophages and predators. Predation on animals representing the detrial food web occurs not only among soil dwelling spiders but at least partially also among those hunting above ground. It can be assumed that spiders which feed on decomposers not only affect the prey population but also indirectly influence the processes of matter decomposition. Two quite different sets of interactions between spiders and the detritus-based food web are potentially relevant. The first set involves cascading, top-down effects of spider predation on rates of nutrient mineralization-spider- initiated trophic cascades in the detrital food web that could affect rates of decomposition and release of nutrients to plants. The second set of interactions relies on a linkage, through spiders, between decomposition and grazing food webs-energy from the detritus food web contributing to elevated spider densities, which in turn might cause lower pest numbers and enhanced net primary production.

Viewing detrital food webs as primarily donor-controlled leads to the conclusion that bottom-up control processes predominate in such webs, i.e., that most linkages of indirect effects are responses to changes in rates of input at the base of the food web. This view is an over-simplification for at least two reasons. First, the rate at which the primary decomposers- the bacteria and fungi, collectively known as the microflora, decompose detritus depends on their population growth rates, which in turn are potentially influenced by their natural enemies (Swift *et al.*, 1979). The second oversimplification is that, ultimately, decomposition processes affect net primary production by altering rates of mineralization, i.e., the rates at which nutrients locked in detritus become available to autotrophs. As a consequence, spiders that feed on detritivores have the potential to influence indirectly the growth of autotrophs by generating trophic cascades in the decomposition food web, thus affecting rates at which nitrogen and other nutrients required by plants enter the nutrient pool. Such linked indirect effects most likely occur in those agroecosystems in which a substantial fraction of litter decomposition occurs on the soil surface, and in which the major consumers of detritus and the microflora are preyed upon by spiders. Ploughing disrupts the litter layer and distributes organic matter throughout the soil; thus, cultivation encourages below-ground, bacterial-based decomposition food webs (Hendrix *et al.*, 1986). The role of spiders in such webs is likely insignificant. Under no-till and conservation tillage conditions, however, litter accumulates above ground, fungi are a major component of the microflora, and microarthropods, primarily mites and Collembola (springtails), are major detritivores/fungivores (Hendrix *et al.*, 1986; Stinner & House, 1990; Robertson *et al.*, 1994). Spiders readily consume Collembola, which constitute a substantial portion of the diet of many species (Hallander, 1970; Schaefer, 1975; Yeargan, 1975; Wingerden, 1975, 1978; Greenstone, 1980, 1983; Nentwig, 1987; Do¨bel & Denno, 1994; Nyffeler *et al.*, 1994).

Spiders have the potential to generate either positive or negative impacts on rates of mineralization, even if one ignores interactions between spiders and other predators. The simplest abstract food chain in a spider-influenced system would consist of three effective trophic levels: detritus, Collembola (realizing that other detritivores play roles similar to Collembola, but Collembola appear to be ubiquitous and abundant), and spiders. The indirect effects of spiders in such a food chain would be to enhance the standing crop of detritus, i.e., to retard the rate of litter decomposition. This model, however, is greatly oversimplified. Fungi play a critical role in decomposing plant litter; and although Collembola consume plant material, many Collembola species are primarily fungivorous (Peterson, 1971; Chen *et al.*, 1996). Thus, a four-level food chain model is more realistic. In grazing food chains of four trophic levels, predators can induce a trophic cascade that negatively impacts the base trophic level-the primary producers. Reasoning by analogy, one might predict that an increase in spider density should lead to a decrease in the amount of detritus, i.e., an increase in decomposition rate, by relieving predation pressure on fungi. However, does an increase in fungal biomass always lead to increased rates of mineralization? Not necessarily, because nutrients can become immobilized in senescent fungal hyphae. Thus, Collembola at intermediate densities can enhance rates of mineralization by consuming senescent fungal hyphae (Parkinson *et al.*, 1977; van der Drift & Jansen, 1977; Warnock *et al.*, 1982; Finlay, 1985; Verhoef & de Goede, 1985; Visser, 1985). Collembola also enhance decomposition by more indirect pathways, i.e. by comminuting the litter (Anderson *et al.*, 1984). Therefore, depression of Collembola populations by spiders could negatively impact rates of litter decomposition and mineralization.

The population build-up of natural enemies is dependent on the availability of suitable host/prey. The abundant detritivores early in the season may be one key to the success of the current rice agroecosystem (Settle *et al.*, 1996). Being polyphagous predators, spiders can prey on alternative prey such as Collembola during fallow periods, hereby maintaining high population levels (Here, alternative prey describes all suitable preys other than the target species). The levels of these alternative prey in turn depend on decaying organic material available in the field. Field and laboratory data from research at the International Rice Research Institute (IRRI) in the Philippines and elsewhere indicates that spiders can survive and build up their populations on alternative prey such as Collembola and dipterans before the crop is established in first weeks after crop establishment (Guo *et al.*, 1995; Settle *et al.*, 1996). Settle *et al.* (1996) were able to increase the number of detritus feeders, such as collembola and of plankton feeders by adding organic material to the rice field in the treated plots. Most interestingly, the number of spiders increased in the same plots. Plankton feeders in that study included mosquito larvae and chironomid midge larvae, of which many species also feed on detritus (Settle *et al.* 1996). In a study at IRRI, the addition of rice straw bundles in the rice field after harvest increased the number of *A. formosana* and *P. pseudoannulata* as well as plant- and leafhoppers (Shepard *et al.*, 1989). Though the study by Shepard *et al.* (1989) did not report effects on Collembola density, high Collembola density can be observed in recently cut straw, so probably the beneficial effect was also due to an increase in Collembola. In upland, rice weed residues placed within the rice fields can significantly increase spider densities (Afun *et al.*, 1999). Apart from providing refuges for predators and increasing the density of alternative prey, organic material will also influence plant nutrition, which in turn can influence herbivores feeding on the crop. One can speculate that this in turn could indirectly affect predators

Food-web interconnections in tropical rice fields, in which detritus is both allo- and autochthonous, constitute a probable example from agricultural systems. Settle *et al.*, (1996) argue that abundant populations of detritivores and planktivores early in the season maintain high densities of generalist predators, which are then able to suppress pest populations on rice later in the season. Field experiments have demonstrated that carabids and spiders can limit aphid numbers in cereal crops (Chiverton 1986; Edwards *et al.*, 1979).

References

Afun, J. V. K., Johnson, D. and Russell-Smith, A. (1999). The effects of weed residue management on pests, pest damage, predators and crop yield in upland rice in Cote d'Ivoire. *Biological Agriculture & Horticulture* **17 (1):** 47-58.

Anderson, J. M., Rayner, A. M. and Walton, D. H. (1984). Invertebrate-microbial interactions. Cambridge Univ. Press, New York.

Chen, B., Snider, R. J. and R. M. Snider (1996) Food consumption by Collembola from northern Michigan deciduous forest *Pedobiologia* **40:** 149–161.

Chiverton, P.A. (1986). Predator density manipulation and its effect on populations of Rhopalosiphum padi (Hom.: Aphididae) in spring barley *Annals of Applied Biology* **109:** 49–60

Dobel, H. G. and Denno, R. F. (1994). Predator-planthopper interactions. Pp. 325–99. In Planthoppers: Their Ecology and Management. (R.F. Denno & T.J. Perfect, eds.). New York: Chapman and Hall.

Drift, van der J, and Jansen, E. (1977). Grazing of springtails on hyphal mats and its influence on fungal growth and respiration *Ecological Bulletins* **25:** 203–09.

Edwards, C. A., Sunderland, K. D. and George, K. S. (1979). Studies on polyphagous predators of cereal aphids. *Journal of Applied Ecology* **16:** 811– 23.

Finlay, R. D. (1985). Interactions between soil microarthropods and endomycorrhizal associations of higher plants. Pp. 319–331. In Ecological Interactions in Soil. Plants, Microbes and Animals. (A.H. Fitter, D. Atkinson, D.J. Read & M.B. Usher, eds). Boston: Blackwell Sci. Publ.

Greenstone, M. H. (1980). Contiguous allotopy of Pardosa ramulosa and Pardosa tuoba (Araneae: Lycosidae) in the San Francisco Bay Region, and its implications for patterns of resource partitioning in the genus. *The American Midland Naturalist journal* **104:** 305– 11.

Guo, Y. J., Wang, N. Y., Jiang, J. W., Chen, J. W. and Tang, J. (1995). Ecological significance of neutral insects as a nutrient bridge for predators in irrigated rice arthropod communities. Chinese *Journal of Biological Control* **11:** 5-9.

Hallander, H. (1970). Prey, cannibalism, and microhabitat selection in the wolf spiders *Oikos* **21:** 337–40.

Hendrix, P. H., Parmelee, R. W., Crossley, D. A., Coleman, D. C., Odum, E. P., Groffman, P. M. (1986). Detritus food webs in conventional and no-tillage agroecosystems. *BioScience* **36:** 374–80.

Nentwig, W. (1987). The prey of spiders. pp. 249- 73. In Ecophysiology of Spiders. (W. Nentwig, ed.). Springer-Verlag, Berlin.

Nyffeler, M., Sterling, W. L. and Dean, D. A. (1994). How spiders make a living. *Environmental Entomology* **23:** 1357–67.

Parkinson, D. S., Visser, S. and Whittaker, J. B. (1977). Effects of collembolan grazing on fungal colonization of leaf litter. *Ecological Bulletins* **25:** 75–9.

Peterson, H. (1971). The nutritional biology of Collembola and its ecological significance. A review of recent literature with a few original observations. *Entomology Meddr* **39:** 97–118.

Riechert, S. E. and Bishop, L. (1990). Prey control by an assemblage of generalist predators: spiders in garden test systems. *Ecology* **71:** 144-50.

Robertson, L. N., Kettle, B. A. and Simpson, G. B. (1994). The influence of tillage practices on soil macrofauna in a semi-arid agroecosystem in northeastern Australia. *Agriculture, Ecosystem and Environment* **48:** 149–156.

Schaefer, M. (1975). Experimental studies on the importance of interspecies competition for the lycosid spiders in a salt marsh. Pp. 86–90. *In Proc. Sixth Intern. Arachnol. Cong., Amsterdam.* (L. Vlijm, ed.).

Settle, W. H., Ariawan, H., Astuti, E. T., Cahyana, W., Hakim, A. L., Hindayana D, Lestari A S and Sartanto P (1996) Managing tropical rice pests through conservation of generalist natural enemies and alternative prey *Ecology* **77:** 1975– 88.

Shepard. B. M., Rapusas, H. R., and Estano, D. B. (1989). Using rice straw bundles to conserve beneficial arthropod communities in the rice field. *International Rice Research Notes* **14**: 30-31.

Stinner, B. R. and House, G. J. (1990). Arthropods and other invertebrates in conservation-tillage agriculture. *Annual Review of Entomology* **35:** 299–318.

Swift, M. J., Heal, O. W., Anderson, J. M. (1979). Decomposition in terrestrial ecosystems. Blackwell Scientific Oxford, UK

Verhoef, H. A. and R, M. Goede (1985). Effects of collembolan grazing on nitrogen dynamics in a coniferous forest. Pp. 367–376 In Ecological Interactions in Soil. Plants, Microbes and Animals. Blackwell Scientific Oxford, UK

Visser, S. (1985). Role of the soil invertebrates in determining the composition of soil microbial communities. Pp. 297–317. In Ecological Interactions in Soil. Plants, Microbes and Animals. Blackwell Scientific Oxford, UK

Warnock, A. J., Fitter, A. H. and Usher, M. B. (1982). The influence of the springtail Folsomia candida (Insecta, Collembola) on the mycorrhizal association of leek Allium Glomus fasciculatus *New Phytologist* **90:** 285–92

Wingerden, W. K. and R. E. van (1978). Population dynamics of Erigone arctica (White) (Araneae, Linyphiidae). II. *Symposia of the Zoology Society of London* **42:** 195–202.

Wingerden, W. K. and van R. E. (1975). Population dynamics of Erigone arctica (White) (Araneae, Linyphiidae). Pp. 71–76. In Proc. XIth Intern. Cong. Arachnol., Amsterdam. (L. Vlijm, ed.).

Yeargan, K. V. (1975). Prey and periodicity of Pardosa ramulosa (McCook) in alfalfa *Environmental Entomology* **4:** 403–08.

Chapter 9

Reduction of Insect Pest Densities by Spiders

Amandeep Singh, Ripudaman Parihar and Randeep Singh

Many studies have demonstrated that spiders can significantly reduce prey densities. Lang *et al.* (1999) found that spiders in a maize crop depressed populations of leafhoppers (Cicadellidae), thrips (Thysanoptera), and aphids (Aphididae). The three most abundant spiders in winter wheat, *Pardosa agrestis* (Westring) and two species of Linyphiidae, reduced aphid populations by 34% to 58% in laboratory studies (Marc *et al.*, 1999). Both web-weaving and hunting spiders limited populations of phytophagous Homoptera, Coleoptera, and Diptera in an old field in Tennessee (Riechert and Lawrence, 1997). Spiders have also proven to be effective predators of herbivorous insects in apple orchards, including the beetle *Anthonomus pomorum* Linnaeus, and Lepidoptera larvae in the family Tortricidae (Marc and Canard, 1997). In no-till corn, wolf spiders (Lycosidae) reduce larval densities of armyworm, *Pseudaletia unipunctata* (Haworth) (Laub and Luna, 1992). Wolf spiders also reduced densities of sucking herbivores (Delphacidae and Cicadellidae) in tropical rice paddies (Fagan *et al.*, 1998). Spiders are capable of reducing populations of herbivores that may not be limited by competition and food availability in some agroecosystems. Several studies have shown that insect populations significantly increase in the absence of spiders. Riechert and Lawrence (1997) reported that plots in an old field from which spiders had been removed had significantly higher herbivorous insect numbers than in those plots that contained spiders.

Agricultural fields that are frequently sprayed with pesticides often also have lower spider populations (Bogya and Markó, 1999; Feber *et al.*, 1998; Holland *et al.*, 2000; Amalin *et al.*, 2001). In general, spiders are more sensitive as compared to many pests to some pesticides, such as the synthetic pyrethroids, cypermethrin and deltamethrin; the organophosphates, dimethoate and malathion; and the carbamate, carbaryl. A decrease in spider populations as a result of pesticide use can result in an outbreak of pest populations (Brown *et al.*, 1983; Birnie *et al.*, 1998; Huusela-Veistola, 1998; Marc *et al.*, 1999; Holland *et al.*, 2000; Tanaka *et al.*, 2000). Spiders can lower insect densities, as well as stabilize populations, by virtue of their top-down effects, microhabitat use, prey selection, polyphagy, functional responses, numerical responses, and obligate predatory feeding strategies.

Nevertheless, as biological control agents, spiders must be present in crop fields and prey upon specific agricultural pests. Spiders may be important mortality agents of crop pests such as aphids, leafhoppers, planthoppers, fleahoppers, and Lepidoptera larvae. However, the same species of spider that feeds on specific type of pests in one location may feed on another type of insects (Table-3).

Table-3. Common Crop pest orders/ families/species and the spiders that are known to prey upon them

S. No.	Pest Families (F)/ Orders (O)/ Species (S)	Common Name	Predatory Spiders
1.	Aphididae (F)	Aphids	Salticidae, Thomisidae, Linyphiidae, *Clubiona* spp., *Pardosa* spp., *O. salticus, E. atra, L. tenuis*
2.	Acrididae (F)	Grasshopper	*R. rabida, P.audax, A,labyrinthica, Argiope* spp
3.	Cicadelidae (F)	Leafhopper	Salticidae, Thomisidae, Thendiidae, *P. pseudoannulata, Pardosa* spp, *O. salticus, P. viridians, P.audax*
4.	Chrysomelidae (F)	Flea beetles	Salticidae, Agelenidae, Araneidae, Thendiidae
5.	Thysanoptera (O)	Thrips	Salticidae, Thendiidae, *Pardosa* spp., *P. audax*
6.	Lepidoptera (O)	Caterpillars	Linyphiidae, *C. mildei, Clubiona* spp., *L. artelucana, Hogna* spp., *O. salticus, P.audax, Misumenops* spp., *A. labyrinthica*
7.	*Lygus lineolaris* (S)	Tamished plant bug	Salticidae, Linyphiidae, *C. inclusum, L. antelucana, Pardosa* spp., *O. salticus, P. audax, P. galathea, Misumenops* spp., *P. mira*
8.	*Schizaphis graminum* (S)	Green bug	*P.audax, P.galathea*
9.	*Bissus leucopterus* (S)	Chinch bug	*C. inclusum, Pardosa* spp., *P.galathea, Misumenops spp., P. mira*
10.	*Spissistilus festinus* (S)	Three corned alfalfa hopper	*C. inclusum, L. antelucana, Pardosa* spp., *O. salticus, P. audax, P. galathea, Misumenops* spp., *P.mira*
11.	*Nilaparvata lugens* (S)	Brown planthopper	*P. pseudoannulata, U. insecticeps*
12.	*Pseudatomoscelis seriatus* (S)	Cotton fleahopper	*O. salticus, P. viridians*
13.	*Empoasca fabae* (S)	Potato leafhopper	*O. salticus, P. audax*

(Source- Maloney, 2003)

(a) Spider Assemblages

Numerous researchers have stressed that an assemblage of spider species is more effective at reducing prey densities than a single species of spider (Greenstone, 1999; Sunderland, 1999). Provencher and Riechert (1994) used computer simulations and field tests to demonstrate that an increase in spider species richness leads to a decrease in prey biomass. Riechert and Lawrence (1997) found that pest population was lower in test plots that contained a sheet-web weaver (*Florinda coccinea* (Hentz)), an orb-web weaver (*Argiope trifasciata* (Forskal)), and two wolf spiders (*Rabidosa rabida* (Walckenaer) and *Pardosa milvina* (Hentz)) than in plots that contained only one of these species. Foraging behavior may be enhanced by the presence of other spider taxa. In agricultural fields in Ohio, the cob-web weaver, *A. tepidariorum* and the orb-web weaver, *Nuctenea cornuta* (Clerck) caught more prey per spider when in groups than when alone. Rypstra (1997) also demonstrated that prey capture also was higher in mixed-species groups than in single-species groups. However, competition among some spiders may limit their efficiency to decrease prey population (Marshall and Rypstra, 1999b). A diverse group of spiders may be effective in biological control because they differ in hunting strategies, habitat preferences, and active periods. Because of the typical diversity of spiders in an agricultural ecosystem, there will probably be one or more species that will attack a given pest (Marc *et al.*, 1999). Different species of spiders feed on different insects at different times of the day, so a loss in community diversity of spiders can result in some prey species being released from predation pressure (Riechert and Lawrence, 1997). Variation in body size of both predator and prey species also contributes to prey reduction, with larger spiders taking larger prey and smaller spiders taking smaller prey (Nentwig and Wissel, 1986; Nyffeler *et al.*, 1994a). In addition to this, larger spiders consume disproportionately more prey than smaller spiders (Provencher and Riechert, 1994).

So, it is important to have an assemblage of different species of spiders rather than just one species so that there will be predators of appropriate size classes and foraging modes to prey upon different prey life stages throughout the growing season. This size class effect can best be accomplished through an assemblage of species because spiders usually have a long generation time compared to their prey (Riechert, 1999). Assemblages of spiders have usually been described within the ecological framework of guilds. However, applying the ecological guild concept to spiders has usually been taxon based instead of resource based. It has been suggested that spider assemblages are the unit of predation pressure exerted by spiders and mathematical modeling has shown that spider communities that naturally exhibit an uneven age-structure and have strong migratory and aggregation tendencies offer the greatest potential pest suppression (Riechert, 1999).

Prey Specialization

Some degree of specialization or monophagy by a predator on prey is assumed to be necessary for the predator to reduce populations of that particular prey. Because of this assumption, spiders, which are polyphagous, generalist predators, were traditionally thought incapable of controlling prey populations (Riechert and Lockley, 1984). However, spiders may be more specialized on particular prey than is often realized. It is common that when spiders have an abundance of prey, they become more selective (Riechert and Harp, 1987). Toft (1999) reported that it might be counter productive for a spider to feed on any prey since some might be toxic or deficient in nutrients. In addition, each species of spider occupies a specific region of the agricultural habitat, from the ground to the top of the canopy. Different prey species can be found in different microhabitats as well. From this, it can be concluded that prey specialization by spiders could be an attribute found in ecosystems, rather than in the laboratory. Temporal differences in prey-capture activities are found among spiders and may lead to specialization of diets. For example, some web weavers are diurnal, spinning their webs during the day; others are nocturnal, spinning and capturing prey at night. Most hunting spiders that rely on visual and vibratory cues are diurnal, but there are exceptions, with some hunters active during night. Spiders, therefore, will only catch prey they encounter during their active period (Marc and Canard, 1997; Marc *et al.*, 1999). For example, in France, nocturnal and diurnal wandering spiders forage on the trunk and in the foliage of apple trees, while ambush species forage among the leaves and flowers. Tubular web species reside under the bark of the trees, while other web weavers occupy different microhabitat between leaves and branches (Marc and Canard, 1997). In addition to microhabitat preferences, spiders also have feeding preferences. Spiders feed predominantly on small-sized prey relative to their own size (prey length ≤ predator length). They usually only eat prey that is 50% to 80% of their size, with web weavers more adept at catching larger prey; smaller prey are typically ignored (Nentwig and Wissel, 1986; Nyffeler *et al.*, 1994a; Marc and Canard, 1997; Marc *et al.*, 1999). Some species of spiders also select insect prey to balance their amino acid requirements (Greenstone, 1979). Although spiders are polyphagous predators, but their hunting strategies and microhabitat preferences cause each species to be specialized (Nyffeler *et al.*, 1994a; Marc and Canard, 1997; Marc et al., 1999; Nyffeler, 1999). Some types of spiders feed on a particular type of prey. The bolas spiders and ladder web spiders (Araneidae) built webs that are specially adapted to catch adult Lepidoptera. Smaller web weavers, such as Linyphiidae and Dictynidae, capture mainly soft-bodied insects such as aphids. Some cobweb weavers (Theridiidae) feed on ants, including fire ants. A number of jumping spider species (Salticidae) are also behaviorally adapted to feed on ants (Nyffeler *et al.*, 1994a; Jackson and Pollard, 1996). Some web-weavers also show similar preferences. Although insects belonging to 17 different orders were observed in webs spun by *Argiope argentata* (Fabricius), 62% of prey consumed by this spider were stingless bees of the genus *Trigona* (Craig and Bernard, 1990). Some web-weaving spiders also preferentially reject prey such as Coleoptera, either ignoring them or cutting them out of the web (Nyffeler *et al.*, 1994a). Indeed, many spiders show behavioral specializations and prey preferences that make them able to effectively limit certain prey populations.

(d) Role of Generalist Spider

Some researchers argue that generalist predators may be more effective than specialists at reducing and stabilizing prey densities (Symondson *et al.*, 2002).

Young and Edwards (1990) studied the spiders in united states field crops and their potential effect on crop pests. The workers suggested that hunting spiders might be better at controlling pests than web-weavers because most species of hunting spiders are capable of capturing a wide variety of prey types and sizes. Web-weaving spiders, however, are more specialized. Despite being capable of capturing grasshoppers and beetles, they usually capture aphids and flies only, and often have little to no impact on plant bugs, weevils, leaf beetles, and caterpillars. Of course, spiders do not consume only pestiferous herbivores. Being generalists, they feed on more than one trophic level in a food web (Morin, 1999). Compared to irruptive species such as insect pests that feed on only one trophic level, some spiders exhibit relatively stable population dynamics (Riechert and Lockley, 1984; Nentwig, 1988). Although model food webs predict that polyphagy will lead to instability. But studies of natural communities show that food chains containing generalists are more stable (Wise, 1999). Predators feeding on multiple prey species in multiple trophic levels are more likely to withstand declines in the abundance of one prey species than predators that specialize on that species (Reichert, 1999). Popuaation abundance of spiders that feed on specific prey fluctuate while polyphagous species are less likely to fluctuate and maintain consistently high populations (Morin, 1999). In agroecosystems, generalist spiders, may maintain populations in periods of low pest population by preying upon other insects, including harmless and beneficial insects (Riechert and Lockley, 1984; Nyffeler *et al.*, 1992, 1994a). Despite the potential to create stable predator populations, polyphagy may be a disadvantage in systems such as agricultural fields, where food chains may be short and simple. In a food chain consisting of three levels- primary predator, herbivore, and producer, the herbivore is not limited by competition but by predation.

References

Amalin, D. M., PeÒa J. E., McSorley, R., Browning, H. W. and Crane, J. H. (2001). Comparison of different sampling methods and effect of pesticide application on spider populations in lime orchards in South Florida *Environmental Entomology* **30:** 1021-27.

Birnie, L., Shaw, K., Pye, B., and Denholm, I. (1998). Considerations with the use of multiple dose bioassays for assessing pesticide effects on non-target arthropods. In Proceedings, Brighton Crop Protection Conference ñ Pests and Diseases. 16-19 November, Brighton, UK. pp. 291-296.

Bogya, S. and MarkÛ, V. (1999). Effect of pest management systems on ground-dwelling spider assemblages in an apple orchard in Hungary *Agriculture, Ecosystem and Environment* **73:** 7-18.

Brown, K. C., Lawton, J. H. and Shires, S. (1983). Effects of insecticides on invertebrate predators and their cereal aphid (Hemiptera: Aphidae) prey: laboratory experiments *Environmental Entomology* **12:** 1747-50.

Craig, C. L. and Bernard, G. D. (1990). Insect attraction to ultravioletreflecting spider webs and web decorations *Ecology* **71:** 616-23.

Fagan, W. F., Hakim, A. L., Ariawan, H. and Yuliyantiningsih, S. (1998). Interactions between biological control efforts and insecticide applications in tropical rice agroecosystems: the potential role of intraguild predation. Biological Control: Theory and Applications in Pest Management. **13:** 121-26.

Feber, R. E., Bell, J., Johnson, P. J., Firbank, L. G. and Donald, D. W. (1998). The effects of organic farming on surface-active spider (Araneae) assemblages in wheat in southern England, UK. *Journal of Arachnology* **26:** 190-202.

Greenstone, M. H. (1979). Spider feeding behavior optimizes dietary essential amino acid composition *Nature* **282:** 501-03.

Greenstone, M. H. and Sunderland, K. D. (1999). Why a symposium on spiders in agroecosystems now *Journal of Arachnology* **27:** 267-69.

Holland, J. M., Winder, L. and Perry, J. N. (2000). The impact of dimethoate on the spatial distribution of beneficial arthropods in winter wheat *Annals of Applied Biology* **136:** 93-105.

Jackson, R. R. and Pollard, S. D. (1996). Predatory behavior of jumping spiders. Annual *Reviwe of Entomology* **41:** 287-308.

Lang, A., Filser, J. and Henschel, J. R. (1999). Predation by ground beetles and wolf spiders on herbivorous insects in a maize crop. *Agriculture Ecosystem and Evironment* **72:** 189-99.

Laub, C. A. and Luna, J. M. (1992). Winter cover crop suppression practices and natural enemies of armyworm (Lepidoptera, Noctuidae) in no-till corn *Environmental Entomology* **21:** 41-9.

Marc, P. and Canard, A. (1997). Maintaining spider biodiversity in agroecosystems as a tool in pest control *Agriculture Ecosystem and Environment* **62:** 229-35.

Marc, P., Canard, A. and Ysnel, F. (1999). Spiders (Araneae) useful for pest limitation and bioindication *Agriculture, Ecosystems and Environment* **74:** 229–73.

Marshall, S. D. and Rypstra, A. L. (1999). Patterns in the distribution of two wolf spiders (Araneae: Lycosidae) in two soybean agroecosystems. *Environmental Entomology* **28:** 1052-59.

Morin, P. J. (1999). Community Ecology. Blackwell Science, Inc., Malden, MA.

Nentwig, W. (1988). Augmentation of beneficial arthropods by stripmanagement: succession of predacious arthropods and long-term change in the ratio of phytophagous and predacious arthropods in a meadow *Oecologia* **76:** 597-606.

Nentwig, W. and Wissel, C. (1986). A comparison of prey lengths among spiders *Oecologia* **68:** 595-600.

Nyffeler, M. (1999). Prey selection of spiders in the field *Journal of Arachnology* **27:** 317-24.

Nyffeler, M., Dean, D. A. and Sterling, W. L. (1992). Diets, feeding specialization, and predatory role of two lynx spiders, Oxyopidae salticus and Peucetia viridans (Araneae: Oxyopidae), in a Texas cotton agroecosystem *Environmental Entomology* **21:** 1457-65.

Nyffeler M, Sterling W L and Dean D A (1994) How spiders make a living *Environmental Entomology* **23:** 1357-67.

Provencher, L. and Riechert, S. E. (1994). Model and field test of prey control effects by spider assemblages Environmental Entomology **23:** 1-17

Riechert, S. and Harp, J. (1987). Nutritional ecology of spiders. In Nutritional Ecology of Insects, Mites, Spiders, and Related Invertebrates, ed. F. Slansky and J.G. Rodriguez. John Wiley & Sons, New York. pp. 645–672.

Riechert, S. E. (1999). The hows and whys of successful pest suppression by spiders: insights from case studies *Jounal of Arachnology* **27:** 387-96.

Riechert, S. E. (1999). The hows and whys of successful pest suppression by spiders: insights from case studies Journal of Arachnology **27:** 387-96.

Riechert, S. E. and Lawrence, K. (1997). Test for predation effects of single versus multiple species of generalist predators: spiders and their insect prey *Entomologia Experimentalis et Applicata* **84:** 147-55.

Riechert, S. E. and Lawrence, K. (1997). Test for predation effects of single versus multiple species of generalist predators: spiders and their insect prey *Entomologia Experimentalis et Applicata* **84:** 147-55.

Riechert, S. E. and Lockley, T. (1984). Spiders as biological control agents Annual Review of Entomology **29:** 299-320.

Rypstra, A. L. (1997). Foraging enhanced by the presence of other predators: prey capture of single- and mixed-species groups of spiders *Bulletin of Ecological Society of America* **78 (4):** 31.

Sunderland, K. D. (1999). Mechanisms underlying the effects of spiders on pest populations *Journal of Arachnology* **27**: 308-16.

Symondson, W. O. C., Sunderland, K. D. and Greenstone, M. H. (2002). Can generalist predators be effective biocontrol agents Annual Review of Entomology **47:** 561-94.

Tanaka, K., Endo, S. and Kazano, H. (2000). Toxicity of insecticides to predators of rice planthoppers: spiders, the mired bug, and the dryinid wasp. Applied Entomology and Zoology **35:** 177-87.

Toft, S. (1999). Prey choice and spider fitness. *J. Arachnol.* **27**:301–7.

Wise, D. H. and Chen, B. (1999). Impact of intraguild predators on survival of a forest-floor wolf spider *Oecologia* 121: 129-37.

Young, O. P., and Edwards, G. B. (1990). Spiders in United States field crops and their potential effect on crop pests *Journal of Arachnology* **18:** 1-27.

Chapter 10

Can Spider Be Effective Biocontrol Agent

Kiranjot Kaur and Amandeep Singh

The importance of spiders as pest control agents is not recent discovery. Their value has been acknowledged by farmers from time immemorial. Spiders are predators of pests like thrips, caterpillars, aphids, plant bugs, leaf hoppers, flies, etc. Conservation and augmentation of spiders in the fields is a simple, yet efficient method of pest control. Once pesticides are kept away from the fields, spiders invariably take shelter in the fields, feed on the pests and add to the productivity.

Spiders must be capable of fulfilling both of pest reduction and pest stabilization requirements. For a predator to effectively and economically control an insect pest, the predator must be capable of maintaining the population of pest below the economic threshold level and also to stabilize those pest densities over time. If the pest population is not stable, the predator may drive the prey to local extinction, then die off itself, thus allowing for the potential of an unchecked secondary pest outbreak in the absence of this predator (Morin, 1999; Pedigo, 2001).

In summary, spiders can be effective predators of herbivorous insect pests, and can exert considerable top-down control, often catching more insects than they actually consume. Despite the potential for competition and intraguild predation, a diverse assemblage of spiders may have the greatest potential for keeping pest densities at low levels. The research on spiders has been focused on wandering spiders, may be because web weavers may either be unable to establish webs or catch pest insects. The spiders that are most efficient at capturing insect pests are those that forage on the plant itself. Spiders show both functional responses and numerical responses to prey densities, although they may not be able to display long-term tracking of any one particular prey species. Due to these density dependent responses, as well as polyphagy in times of low pest levels, spider populations in agroecosystems are stable and can be maintained at low levels even in the absence of pest. Spiders are excellent biological pest management agents because they exhibit the ability to lower and stabilize the pest population. In rice fields in Asia, however, spiders are often purposefully introduced into fields. In China, farmers build straw or bamboo shelters for spiders and then move these shelters to paddy fields which are experiencing pest outbreaks. This method of spider augmentation led to a 60% reduction in pesticide use (Riechert and Bishop, 1990; Marc *et al.*, 1999). In Japan, spider populations are maintained and enhanced by the release of *Drosophila* fruit flies into fields when pest insects are not abundant (Marc *et al.*, 1999). Grounddwelling spiders such as lycosids are one of the most important predators of leafhopper and planthopper pests of rice, and the addition of wolf spiders to rice paddies can result in reductions in pest populations similar to that seen with insecticide use (Nyffeler and Benz, 1987; Fagan *et al.*, 1998; Geetha and Gopalan, 1999; Jalaluddin *et al.*, 2000)

References

Fagan, W. F., Hakim, A. L., Ariawan, H. and Yuliyantiningsih, S. (1998). Interactions between biological control efforts and insecticide applications in tropical rice agroecosystems: the potential role of intraguild predation. Biological Control: Theory and Applications in Pest Management **13:** 121-26.

Geetha, N. and Gopalan, M. (1999). Effect of interaction of predators on the mortality of nymphs of brown plant hopper, Nilaparvata lugens Stal *Journal of Entomological Research* **23:** 179-81.

Jalaluddin, S. M., Mohan, R., Velusamy, R., and Sadakathulla, S. (2000). Predatory behaviour in rice varieties under sodic soil conditions. *Entomology journal* **25:** 249-347.

Marc, P., Canard, A. and Ysnel, F. (1999). Spiders (Araneae) useful for pest limitation and bioindication. *Agriculture, Ecosystem and Environment* **74:** 229-73.

Morin, P. J. (1999). Community Ecology. Blackwell Science, Inc., Malden, MA.

Nyffeler, M. and Benz, G. (1987). Spiders in natural pest control: a review *Journal of Applied Entomology* **103:** 321-39.

Pedigo, L. P. (2001) Entomology and Pest Management, 4th ed. Prentice Hall, New Jersey.

Riechert, S. E. and Bishop, L. (1990). Prey control by an assemblage of generalist predators: spiders in garden test systems *Ecology* **71:** 1441-50.

Chapter 11

Effect of Pesticide on Spiders

Amandeep Singh, Jyoti Mahajan and Randeep Singh

Synthetic Pesticides

Many farmers use chemical pesticides to control insect pests which also have adverse affects on useful insects and other organisms. To preserve the spiders, it is neccessory to regulate the use of pesticides. Therefore, an ideal biological control agent, would be one that is least affected by the use of synthetic insecticides. Although spiders may be more sensitive to insecticides than insects due to their relatively long-life spans, however, some spiders show tolerance, perhaps even resistance, to some pesticides. Spiders are less affected by fungicides and herbicides than by insecticides (Yardim and Edwards, 1998). Spiders such as the wolf spiders *P. pseudoannulata* are not affected by botanical insecticides such as Neem-based chemicals (Theiling and Croft, 1988; Markandeya and Divakar, 1999). They are also generally more tolerant of organophosphates and carbamates than of pyrethroids, organochlorines and various acaricides, although this tolerance may be due to genetic resistance bred over a period of continuous exposure (Theiling and Croft, 1988; Wisniewska and Prokopy, 1997; Yardim and Edwards, 1998; Marc *et al.*, 1999; Tanaka *et al.*, 2000). For example, *P. pseudoannulata* (Lycosidae), *Tetragnatha maxillosa* Thorell (Tetragnathidae), *Ummeliata insecticeps* (Bösenberg et Strand) and *Gnathonarium exsiccatum* (Wider) (Linyphiidae) were highly sensitive to the pyrethroid deltamethrin, but very tolerant of the organophosphate diazinon and the carbamate carbaryl (Tanaka *et al.*, 2000). Rodrigues *et al.* (2013) worked on spider guilds in insecticide-treated rice and reported that web weavers were often negatively affected. Lee *et al.*, (1993) recorded higher numbers of hunters than web weavers after insecticide application in South Korea. Haughton *et al.* (1999) documented stronger effects on weavers both after insecticide and herbicide application in the UK.

Although we have also been able to show that weavers decrease more than hunters, in our case dividing this guild in two reveals a stonger insecticide effect specifically on orbicular web builders (mainly represented by Araneidae and Tetragnathidae). A strong negative response by web weavers to insecticide is expected: both spiders and webs can be contaminated with the insecticide as the webs are more exposed, to increase the likelihood of capturing prey. On the other hand, hunters are much more mobile, being able to recolonise sprayed areas faster. However, space web builders reveal a lower impact of the insecticide than orbicular web builders, actually being more abundant in rice with insecticide. Linyphiidae spiders contributed more for this guild in rice with insecticide application. The most probable explanation is that spiders in this family show high dispersal rates due to ballooning (Thomas and Jepson, 1999). Thus, linyphiids can also recolonise insecticide sprayed fields, faster than orb weavers, as do hunters. Some broad-spectrum organophosphates are highly toxic to spiders. For example, dimethoate sprays resulted in 100% mortality to the lycosid *Trochosa ruricola* (De Geer) at concentrations below recommended field application rates (Birnie *et al.*, 1998). The organophosphate, methyl parathion and the pyrethroid, cypermethrin, are highly toxic to spiders in the genus *Erigone* (Linyphiidae), while the carbamate, pirimicarb, is almost harmless (Brown *et al.*, 1983; Huusela-Veistola, 1998). Toft and Jensen (1989) found that sublethal doses of dimethoate and cypermethrin had no effect on development and predation rates of the wolf spider *Pardosa amentata* (Clerck). Very low doses of cypermethrin, cause increased mortality rate of the adult and penultimate females. The insecticides did have knockdown effects that, although not, influence the survival rate in the laboratory, but would likely result in death in the field due to desiccation or predation (Toft and Jensen, 1998).

Table 4. Relative toxicity of insecticides to spider *Pirata subpiraticus*, and the brown planthoppers, *Nilaparvata lugens*

Insecticides	Formulation	Number tested	Mortality of Spider (*P. subpiraticus*) (In %)	Mortality of Brown Planthopper (*N. lugens*)
BPMC	Emulsion	60	86.7	100
BPMC	Dust	60	100	100
Carbofuran	Granules	60	100	100
Tebufenozide + BPMC	Wettable powder	60	70	100
Bufrofezine + BPMC	Spray	60	10	100
Bufrofezine+ Isoprocarb	Wettable powder	60	10	100
Pyridaphenthio+ metolcarb	Dust	60	100	100
Deltamethrine	Emulsion	60	96.7	100
Tebufenozide+ Bufrofezine	Wettable powder	60	23.3	46.7
Imidacloprid	Spray	60	16.7	100

(Source- Lee and Kim, 2001)

Other factors that influence effects of pesticides on spiders includes type of solvent, soil type, moisture, percent organic matter, temperature, and time of day of spraying. Further, the microhabitat, hunting style, prey preference, and behavior of the spider also influences their response to pesticide application (Marc *et al.*, 1999). Wisniewska and Prokopy (1997) reported increase in spider population, when pesticides were used early in the growing season only. However, spiders recolonize the field when pesticide use ceases after early June. Spatial limitation of pesticides (such as only applying the pesticides to certain plants or certain plots) also result in higher spider population. Since spiders can move out of the treated areas and return when the chemicals dissipate (Riechert and Lockley 1984; Balança and de Visscher, 1997).

The relative toxicity of various insecticides used against the brown planthopper to spiders was tested in the laboratory (Table-4). Carbofuran, widely used in rice fields, is very toxic to spiders (Bae *et al.*, 1994).

Plate-36: Spider, *Pirata subpiraticus* (left) and BPH, *Nilaparvata lugens* (right)

Botanical Pesticide

The impact of synthetic pesticides on beneficial arthropods and the human health risks posed by exposure to these chemicals are issues of growing concern. This has prompted new compounds with reduced environmental persistence and low mammalian and avian toxicity but a fairly broad spectrum of insecticidal activity (Harris, 2000). Chemical pesticides like triazophos (0.05%) and quinalphos (0.05%) caused 64.78 and 46.79% mortality of spiders respectively (Joseph *et al.*, 2010). On the other hand, biological insecticides are least harmful to natural enemies. The study was conducted to compare the effects of different biological insecticides: extracts of *Azadirachta indica* and *Eucalyptus globulus*, and Spinosad on population of spiders in a rice field at an agricultural farm in tehsil Daska of district Sialkot-51310, Pakistan. The study showed the spinosad caused 42.18% mortality followed by Neem (36.68%) and Eucalyptus (33.38%). When aphids treated with spinosad were fed to coccinelid, no predator mortality was recorded. Similarly, Samiayyan and Chandrasekharan (1998) reported that biological insecticides caused less mortality. The percentage reduction of spider population reduced showing reducing trend of toxicity with passage of time. When larvae of *Chrysoperla carnia* were exposed to spinosad 19% mortality after 12 days was observed (Cisneros *et al.*, 2002). Joseph *et al.* (2010) observed 24.50% reduction in spider population when treated with Azadirachtin significantly which was lower than synthetic pesticides. The biological insecticides especially of botanical origin are less harmful and can be used in rice field for pest management without causing adverse effects on natural enemies and environment. The use of inexpensive botanical insecticide will also encourage agroforestry at farm level.

Plate-37. ***Azadirachta indica*** **(left) and** ***Eucalyptus globules*** **(right)**

(Source: www.pureplantessentials.com/products/eucalyptus-eucalyptus-globulus-australia-steam-distilled-leaf, www.ebay.in/itm/Live-Neem-Azadirachta-Indica-Plant-1-Plant-/222623939980)

Compatibility of Spider and insecticides

Most studies on the effect of insecticides on natural enemies in the paddy field have focused on spiders. Many studies have shown that insecticides have a negative effect on the population densities of rice field spiders (Kuno, 1968; Kuno and Hokyo, 1970; Choi *et al.*, 1978; Paik *et al.*, 1979; Kim *et al.*, 1984; Paik and Hwang, 1990; Lee *et al.*, 1993b and Bae *et al.*, 1994). Choi *et al.*, (1978) have suggested that the root zone placement of Carbofuran reduces the density of spiders in rice fields by direct lethal action, and by poisoning their food chain. Most lethal insecticides to spiders are dust or granule types. Many of tested insecticides such as sprays and wettable powders, were selective e.g. *P. subpiraticus* had a higher survival rate after a prolonged exposure. This may be due to the active behavior of this species. It is recommended that if there is no difference in the effect of different types of insecticides on brown planthoppers, the use of dust or granule type insecticides should be avoided.

Several studies have documented that the application of insecticides reduces the population density of spiders in rice fields (Park *et al.*, 1972; Kawahara *et al.*, 1971; Paik *et al.*, 1979; Kim, 1992 and Yun, 1997). They have a negative effect, not only on spider population, but also on species diversity (Lee *et al.*, 1993a, 1993b). The insecticides which generally protect natural enemies often have a negative effect on spiders (Clausen, 1990). Hunting species tend to suffer more damage than web builders (Specht and Dondale, 1960; Legner and Oatman, 1964; Bostanian *et al.*, 1984), and active hunters suffer more as compared to less active (Bostanian *et al.*, 1984). Lee and Kim reported that when insecticides were applied to rice fields, the pest population was decreased. However, the population density of spiders was not affected (Fig.-4). This implies that many of the insecticides used in rice fields have selectivity level, and rice field spiders may have acquired insecticide resistance, as other arthropods have done. Therefore, the reduction of spider populations in rice fields after insecticide applications as shown in previous studies was perhaps not because the spiders were killed by the insecticides, but because their prey were killed. Lee *et al.* (1993a) showed that the density of hunters was higher than that of web builders after insecticides were applied. The workers suggested that during the recovery period, the active hunters played an important role in rebuilding spider densities at this time. Table-8 shows the result of various combinations of insecticides, brown planthoppers and *P. subpiraticus*. The combination of an insecticide and *P. subpiraticus* resulted in the maximum mortality (66.80%) of BPH. When insecticide was used alone, the mortality was 63.4%, while *P. subpiraticus* used alone caused 51.6% mortality. This result suggests that there is a positive relationship between insecticide applications and control by spiders. When 16 individuals of *P. subpiraticus* were inoculated, the mortality was 4.5% higher than when 8 individuals were inoculated. This implies that when population density of spiders is high enough, their potential to control brown planthopper can be as effective as that of insecticides, or even more effective. The control effect of *P. subpiraticus* varied according to sequence of introduction. When brown planthoppers were introduced first, the control value of *P. subpiraticus* was 51.6%. If *P. subpiraticus* was introduced before the planthoppers, the control value was 53.3%.

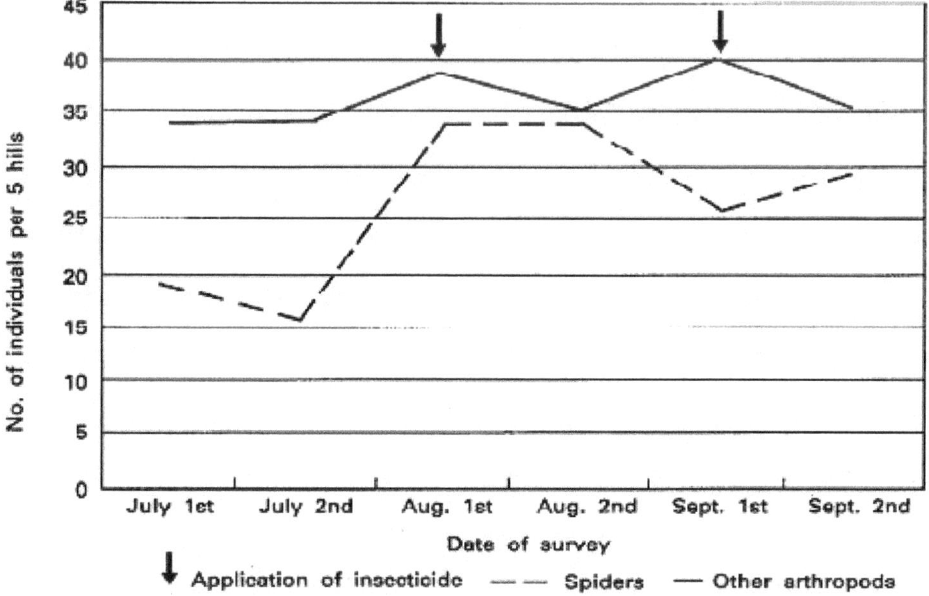

Fig.-4: Seasonal fluctuation in rice field spiders and other arthropods after insecticide applications in Korea, 1996.

(Source-Lee and Kim2001)

Therefore, if the density of rice field spiders is high enough before the immigration of brown planthoppers, the spiders can control the planthoppers effectively and the number and quantity of insecticide applications can be reduced Fig-18. Wide range of spider species inhabit agricultural fields. Their presence limits the habitats open to insect pests. Spiders threaten insect pests with various foraging strategies. Apart from consuming large numbers of insect pests as prey, they have the potential of killing all insects living in their territory. For this reason, spiders are a favorable biological control agents in the agricultural ecosystem. Fagan *et al.* (1998) indicated that treatments which combine the augmentation of natural enemies with insecticide applications may be counterproductive. However, spiders still play an important role in reducing the numbers of insect pests in agricultural fields, even when insecticides are used. In fact, spiders may be responsible for a significant proportion of insect mortality which were thought to be from insecticide applications. A quantitative analysis of the capacity of spiders to suppress insect pests, including the spatial distribution of major species of spiders and pests, should be carried out in the field on a large scale, so that spiders can be successfully used as biological control agents. Other ecological and biological characteristics of spiders also need to be understood. Chiu (1979) suggested that it takes longer for spiders to rebuild their population densities after the application of insecticides than planthoppers and leafhoppers, because spiders have a longer generation interval. Also, the development of selective insecticides, the effect of insecticides on spiders, and appropriate timing and quantities of insecticide applications, should all be considered.

Table-5. Control of brown planthopper (BPH), *Nilaparvata lugens,* after different treatments

Treatment	Inoculation density of brown plant hopper (BPH)	Inoculation density of *P. subpiracticus*	Mean density of BPH after harvest per 8 hills	Mortality of Brown Planthopper (*N. lugens*)
BPH (control)	16	0	128.6	100
BPH + Insecticides	16	0	47.1	100
BPH + Spiders + Insecticides	16	8	42.7	100
BPH + Spiders (Preinoculation of BPH)	16	8	62.3	100

(Source- Lee and Kim 2001)

References

Bae, Y. H., Lee, J. H. and Hyun, J. S. (1994). A Systematic Application of Insecticides to Manage Early Season Insect Pests and Migratory Planthoppers on Rice. *Korean Journal of Applied Entomology* **33(4):** 207-15.

Balanca, G. & Visscher, M. N. (1997). Impacts on nontarget insects of a new insecticide compound used against the Desert Locust (Schistocercagregaria). *Archives of Environmental Contamination and Toxicology* **32:** 58-62.

Birnie, L., Shaw, K., Pye, B. and Denholm, I. (1998). Considerations with the use of multiple dose bioassays for assessing pesticide effects on non-target arthropods. In Proceedings, Brighton Crop Protection Conference ñ Pests and Diseases. 16-19 November, Brighton, UK. pp. 291-96.

Bostanian, N. J., Dondale, C. D., Binns, M. R. and Pitre, D. (1984). Effects of Pesticide on Spiders (Araneae) in Quebec Apple Orchards *Canadian Entomologist* **116:** 663-75.

Brown, K. C., Lawton, J. H. and Shires, S. (1983). Effects of insecticides on invertebrate predators and their cereal aphid (Hemiptera: Aphidae) prey: laboratory experiments. *Environmental Entomology* **12:** 1747-750.

Chiu, S. C. (1979). Biological control of the brown planthopper. In: Brown Plant-hopper, Threat to Rice Production in Asia. International Rice Research Institute, Los Baños, Laguna, Philippines, pp. 335-55.

Choi, S. Y., Lee, H. R. and Ryu, J. K. (1978). Effects of Carbofuran Root-zone Placement on the Spider Populations in the Paddy Fields. *Korean Journal of Plant Protection* **17(2):** 99-103.

Cisneros, J., Goulson, D., Derwent, L. C. and Williams, T. (2002) Toxic effects of spinosad on predatory insects. *Journal of Biological Control* **(23):** 156-63.

Clausen, I.H.S. 1990. Design of research work based on a pilot study dealing with the effect of pesticides on spiders in a sugar-beet field. *Acta Zool. Fennica* 190: 69-74.

Fagan, W. F., Hakim, A. L., Ariawan, H. and Yuliyantiningsih, S. (1998). Interactions between Biological Control Efforts and Insecticide Applications in Tropical Rice Agroecosystems: The Potential Role of Intraguild Predation *Biological Control* **13:** 121-26.

Harris, J. G. (2000). Chemical Pesticides Markets, Health Risks and Residues. CABI, Wallingford, UK.

Haughton, A. J., Bell, J. R., Boatman, N. D. and Wilcox, A. (2001). The effect of herbicide glyphosate on non-target spiders. II. Indirect effects on *Lepthyphantes tenuis* in field margins *Pest Management Science* **57:** 1037-042.

Huusela-Veistola, E. (1998). Effects of perennial grass strips on spiders (Araneae) in cereal fields and impact on pesticide side-effects. *Journal of Applied Entomology* **122:** 575-83.

Joseph, R. A., Premila, K. S. and Mohan, S. S. (2010). Safety of neem products to tetragnathid spiders in rice ecosystem. *Journal of Biopesticide* **3(1):** 88-89.

Kawahara, S., Kiritani, K. and Sasaba, T. (1971). The selective activity of rice-pest insecticides against the green leafhopper and spiders. *Botyu-Kagaku* **36:** 121-28.

Kim, H. S. (1992). Suppressive Effects of Wolf Spider, Pirata subpiraticus (Araneae: Lycosidae) on the Population Density of Brown Planthopper (Nilaparvata lugens Stål). Unpublished Ph.D. Thesis, Dongkuk Univ, Seoul, Korea. 69pp.

Kim, J. B., Cho, D. J., Kim, C. H., Uhm, K. B. and Chang, S. D. (1984). Studies on the seasonal fluctuation and the timing of control for the brown planthopper (Nilaparvata lugens Stål) and white-backed planthopper (Sogatella furcifera Horvath) in southern rice cultural area. Res. Reports, ORD (Office of Rural Development, Korea) **26(2):** 16-20.

Kuno, E. (1968). Studies on the population dynamics of rice leafhopper in a paddy field *Bull. Kyushu Agri. Exp. Stn.* 14(2): 131-246.

Kuno, E. and Hokyo, N. (1970). Comparative analysis of the population dynamics of rice leafhopper, Nephotettix cincticeps Uhler and Nilaparvata lugens Stål, with special reference to natural regulation of their numbers *Researches on Population Ecology* **12:** 154-184.

Lee, H. P., Kim, J. P. & Jun, J. R. (1993). Utilization of insect natural enemies and spiders for the biological control in rice paddy filed, community structure of insect pest and spiders, suppress effect on insect pest by natural enemies, and their overwintering habitats in rice paddy field. *Journal of Agriculture Science* **35:** 261-274.

Lee, H. P., Kim, J. P. and Jun, J. R. (1993). Influences of the insecticidal application on the natural enemies and spider community. *Journal of Industrial Technology Grad* **1:** 295-307.

Legner, E. F. and Oatman, E. R. (1964). Spiders on apple in Wisconsin and their abundance in a natural and two artificial environments. *Canadian Entomologist* **96:** 1202-1207.

Marc, P., Canard, A. and Ysnel, F. (1999). Spiders (Araneae) useful for pest limitation and bioindication. *Agriculture, Ecosystem and Environment* **74:** 229-73.

Markandeya, V. and Divakar, B. J. (1999). Effect of a neem formulation on four bioagents. *Plant Protection Bulletin* **51:** 28-29.

Paik, J. C. and Hwang, J. Y. (1990). Preliminary survey of Natural Enemies of Rice Pests for the Development of Integrated Control Programs. Res. Report RDA. **33:** 153-71.

Paik, W. H. and Namkung, J. (1979). Studies on the Rice Paddy Spiders from Korea. Seoul National University, Korea 101 pp.

Park, J. S., Lee, S. C., Lee, B. H., Kim, Y. I., Park, K. T. and Ahn, K. J. (1972). Influences of the insecticides on the rice insect pests. Res. Report R.D.A.: 146-69.

Riechert, S. E. and Lockley, T. (1984). Spiders as biological control agents *Annual Review of Entomology* **29:** 299-320.

Ryu, J. K., Choi, S. Y., Lee, H. R. and Song, Y. H. (1977). Root-zone placement of carbofuran for control of rice insect pests. *Korean Journal of Plant Protection* **16(4):** 217-20.

Specht, H. B. and Dondale, C. D. (1960). Spider populations in New Jersey apple orchards. *Journal of Economic Entomology* **53:** 810-14.

Tanaka, K., Endo, S. and Kazano, H. (2000). Toxicity of insecticides to predators of rice planthoppers: spiders, the mired bug, and the dryinid wasp. *Applied Entomology and Zoology* **35:** 177-87.

Theiling, K. M. and Croft, B. A. (1988). Pesticide side-effects on arthropod natural enemies: a database summary. *Agriculture, Ecosystem and Environment* **21:** 191-218.

Thomas, C. F. G. and Jepson, P. C. (1999). Differential aerial dispersal of linyphiid spiders from a Grass and a cereal field. *Journal of Arachnology* **27:** 294-300.

Toft, S. and Jensen, A. P. (1998). No negative sublethal effects of two insecticides on prey capture and development of a spider. *Journal of Pesticide Science* **52**: 223-28.

Wisniewska, J. and Prokopy, R. J. (1997). Pesticide effect on faunal composition, abundance, and body length of spiders (Araneae) in apple orchards. *Environmental Entomology* **26:** 763-76.

Yardim, E. N. and Edwards, C. A. (1998). The influence of chemical management of pests, diseases and weeds on pest and predatory arthropods associated with tomatoes. *Agriclture, Ecosystem and Environment* **70:** 31-48.

Yoon, J. C. (1997). Arthropod Community Structure and Changing Patterns in Rice Ecosystems of Korea. Unpublished Ph.D. Thesis, Seoul National University, Seoul, Korea, 105 pp.

Index

Plate-1. Sources: 1. (www.pestnet.org), 2. (Hafiz M 2012), 3. (www.cabbsouat.org), 4. (www.inaturalist.org), 5. (http://blog.naver.com), 6. (Ashley Bradford, 2012), 7, 8 & 9. (www.cabbsouat.org) (**p. 9**).

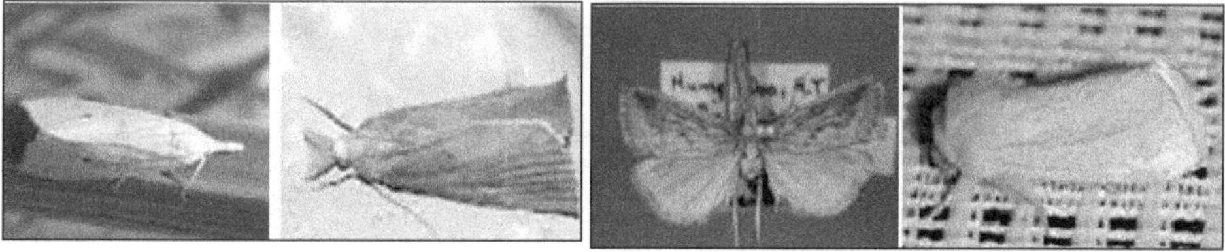

Plate-2. Source: 10, 11, 12 & 13. (www.cabbsouat.org) **(p. 10)**

Plate-3. 14-16. (www.cabbsouat.org), 17. (www.knowledgebank.irri.org), 18-20. (www.cabbsouat.org), 21. (pbt.padil.gov.au) **(p.11)**

Plate-19: *Lycosa* spp. *Spiders*

Plate-20: *Pardosa* spp. spiders

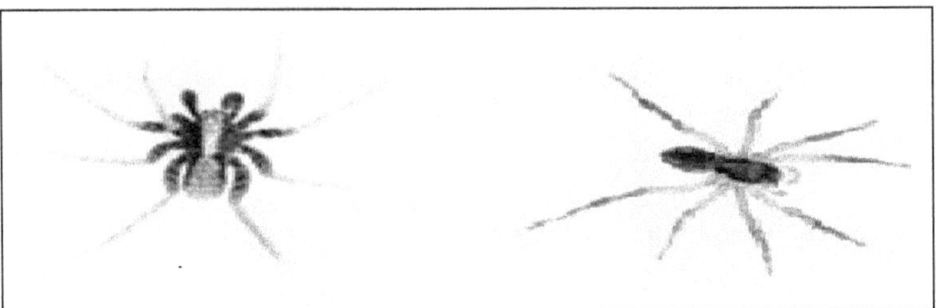

Plate-21: *Wadicosa* spp. Spiders **(p. 38)**

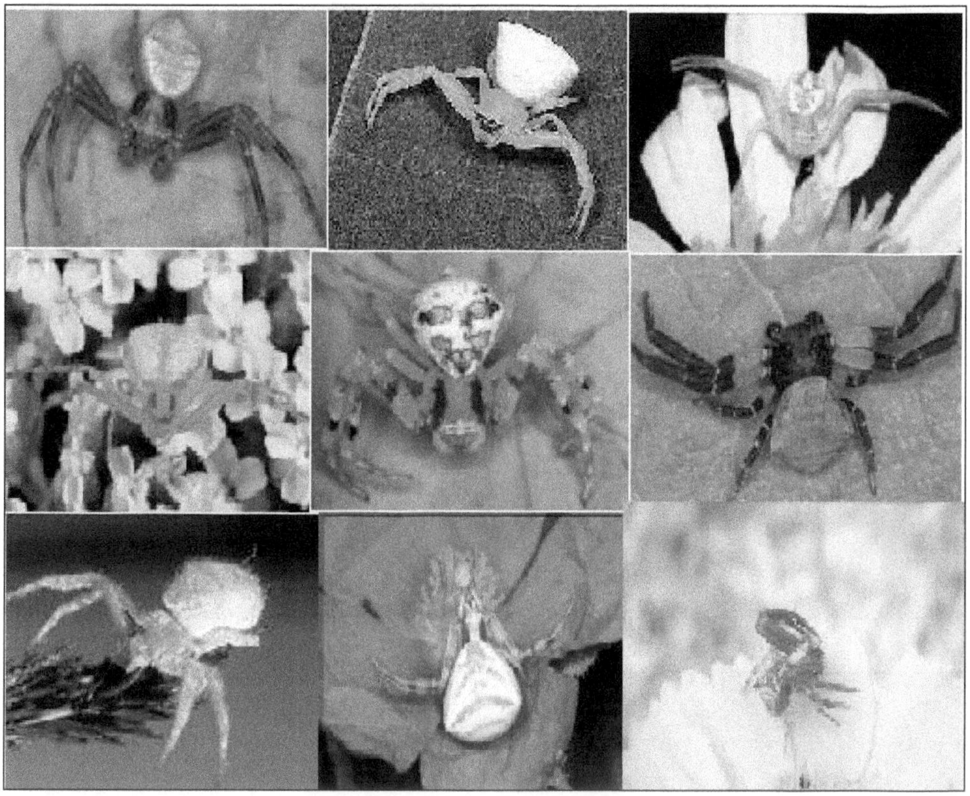

Plate-22: *Thomisus* spp. spiders

Plate-23: *Tetragnata* spp. Spiders

Plate-24: *Argiope* spp. spiders (p. 39)

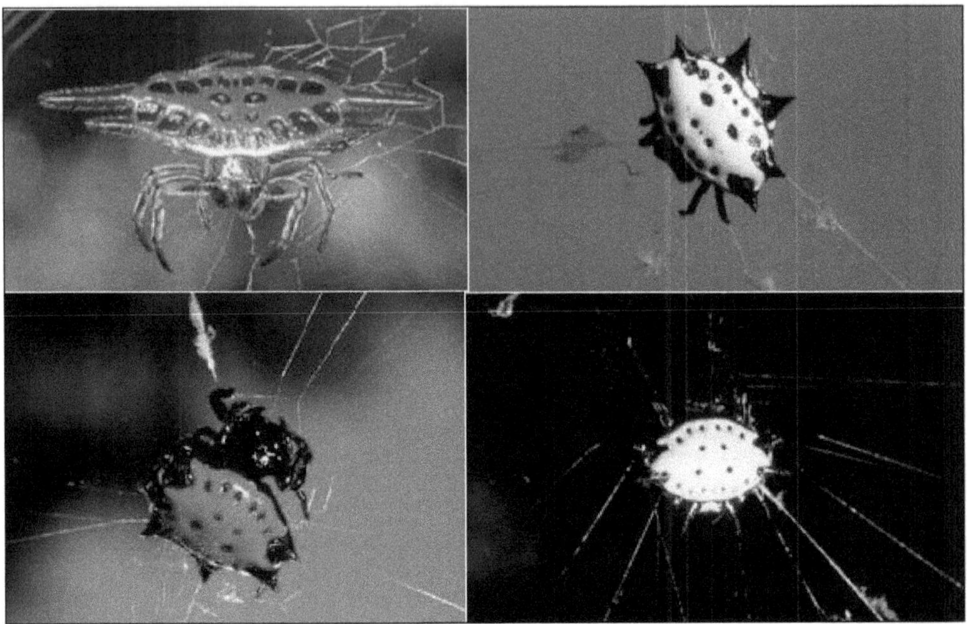

Plate-25: *Gasteracantha* spp. spiders

Plate-26. ***Pholicus*** spp. spiders

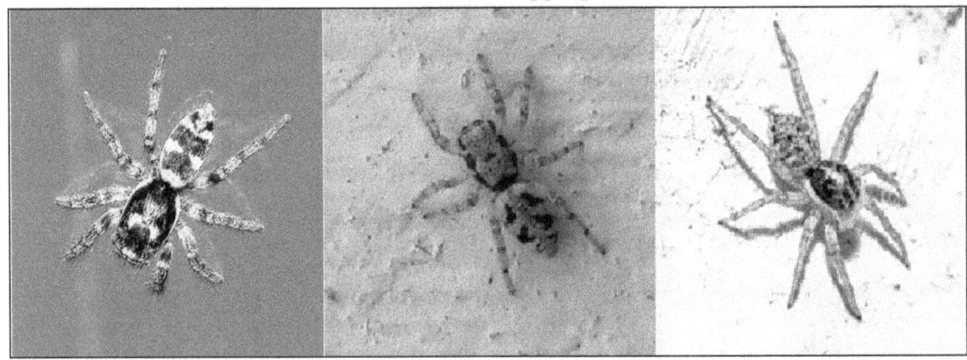

Plate-27. *Salticus* spp. spiders (p. 40)

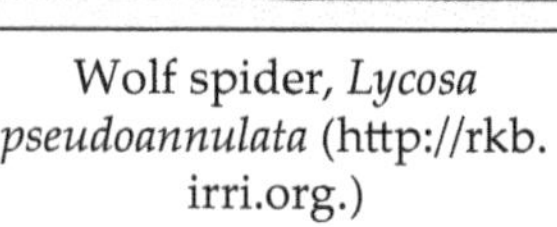

Wolf spider, *Lycosa pseudoannulata* (http://rkb. irri.org.)

(p. 42)

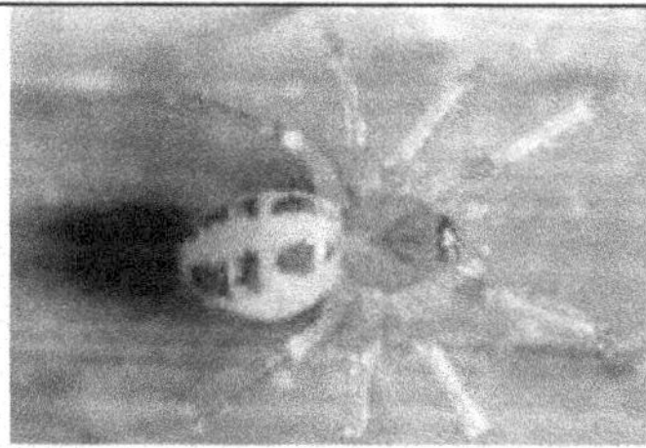

Dwarf spider, *Anyphena formosana* (htpp://rkb.irri.org.)

(p. 42)

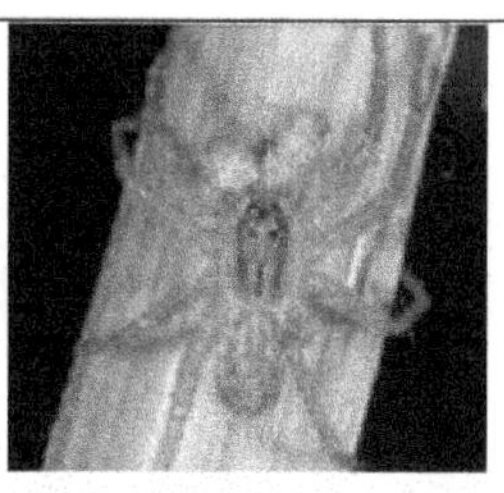

Lycosid spider, *Pardosa pseudoannulata* feeding on BPH (http://rkb.irri.org)

(p. 42)

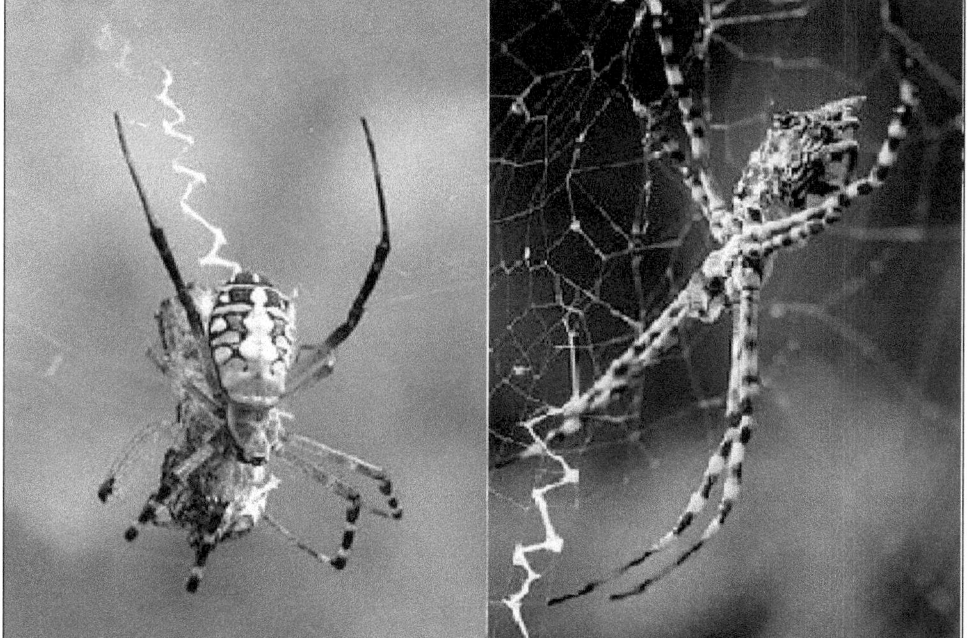

Plate-29: Orb weaver spider **(p.43)**

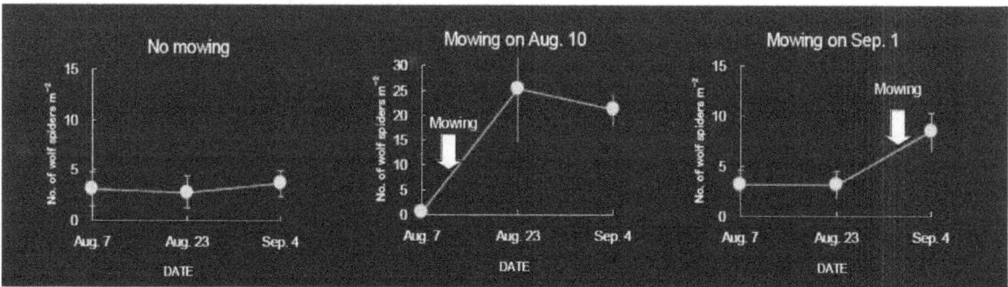

Fig. 3: Effect of the time of mowing on the number of wolf spiders in rice paddy field Error bars indicate standard deviation *(Source- Inagak et al., 2012)* **(p.64)**

Plate-30: Diversity of spiders in rice field **(p.48)**

Plate-31: Straw bundles from sorghum to rice field **(p. 58)**

Plate-34: Laboratory rearing of drosophila for spiders **(p. 63)**

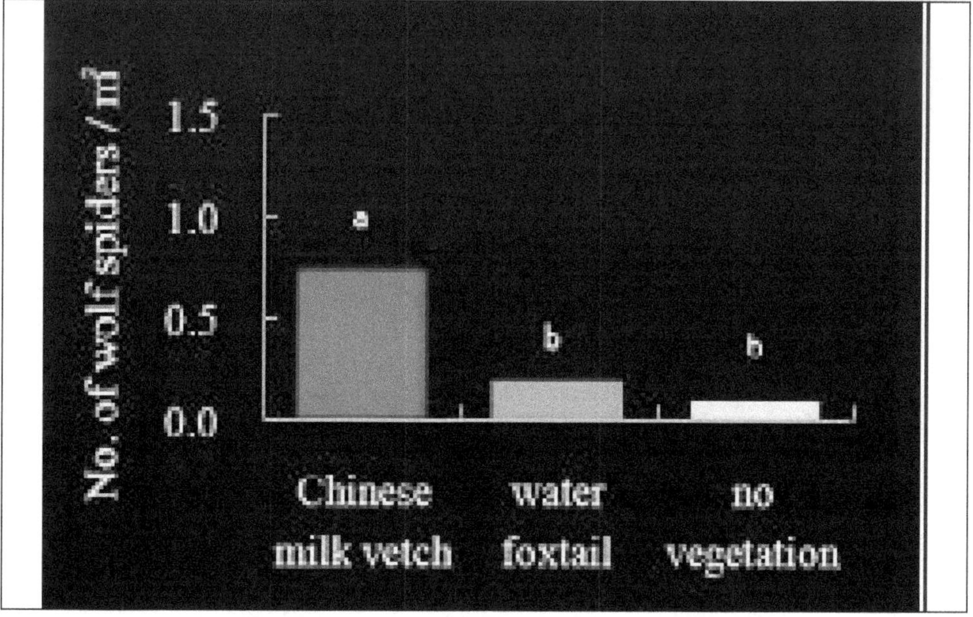

Fig. 4: The number of wolf spiders in rice paddy fields before rice transplanting (Tukey's at p=0.05) *(Source- Inagak et al., 2012)* ***(p. 65)***

Plate-35: Chinese milk verch (left) and water foxtail (right) **(p. 65)**

Plate-36: Spider, *Pirata subpiraticus* (left) and BPH, *Nilaparvata lugens* (right) **(p. 87)**

Plate-37. *Azadirachta indica* **(left) and** *Eucalyptus globules* **(right) (p. 88)**

www.ingramcontent.com/pod-product-compliance
Ingram Content Group UK Ltd.
Pitfield, Milton Keynes, MK11 3LW, UK
UKHW021955270726
14060UKWH00002B/522

9 789390 371990